ЖИДКОМЕТАЛЛИЧЕСКИЕ ТЕПЛОНОСИТЕЛИ

Zhidkometallicheskie Teplonositeli

Liquid-Metal Heat Transfer Media

LIQUID-METAL HEAT TRANSFER MEDIA

S. S. Kutateladze, V. M. Borishanskii,
I. I. Novikov and O. S. Fedynskii

TRANSLATED FROM RUSSIAN

Supplement No. 2
of the Soviet Journal of Atomic Energy
ATOMNAIA ENERGIIA
Atomic Press, Moscow, 1958

Springer Science+Business Media, LLC
1959

ISBN 978-1-4899-4904-2 ISBN 978-1-4899-4902-8 (eBook)
DOI 10.1007/978-1-4899-4902-8

Library of Congress Catalog Card Number: 59-9228

Originally published by Consultants Bureau, Inc. in 1959.

Softcover reprint of the hardcover 1st edition 1959

CONTENTS

FOREWORD

This supplement to the journal of Atomic Energy for 1958 is devoted to problems of utilizing liquid metal heat transfer media in nuclear power. The copious data on the study of heat removal by liquid metals obtained during the past 10 years in the USSR as well as abroad in connection with problems of nuclear power is systemized and generalized in this book. Although the book does not pretend to be an exhaustive presentation, it will be of considerable assistance to scientific workers and engineers who work in the field of reactor design and nuclear power and also in other fields of technology where liquid metal coolants can be utilized.

The major part of the book was written by S. S. Kutateladze, V. M. Borishanskii, and I. I. Novikov; Chapters I, III, V, VIII were written in cooperation with O. S. Fedynskii, G. M. Liamkin, N. A. Prikhodchenko, and Yu. I. Koriakin participated in the preparation of the manuscript.

Chapter 1

BASIC PROPERTIES OF LIQUID METALS

The Structure of Liquid Metals

X-ray and neutron-graphic studies of liquid metals show that the packing of the atoms in a liquid metal is connected in a definite manner with the crystal structure, i.e., with the spatial distribution of the atoms of the given metal in the solid state. For example, for liquid sodium, tin, and other metals the kind of packing is preserved not only at temperatures near the melting point, but also at temperatures which considerably exceed it. The structure of the substance is characterized by the so-called mean coordination number, or the mean number of atoms found on the surface of a sphere whose center is the center of any atom and whose radius is equal to the distance to a neighboring atom. In going over from the crystal state to the liquid, i.e., after melting, the coordination number of all metals, as evident from the data given below, either increases or remains unchanged.

Mercury. With respect to its thermodynamic properties mercury is almost a perfect liquid, i.e., it has densely packed atoms (from the data of an x-ray investigation at high temperatures the coordination number of mercury is equal to 12). With a decrease in temperature the coordination number diminishes to 8-10, while at temperatures near the melting point it becomes equal to 6. From this it follows that near the melting point the structure of mercury in the liquid state is the same as in the crystalline state (simple cubic lattice).

Sodium. In the crystalline state sodium has a volume-centered lattice with a coordination number 8. X-ray investigations of liquid sodium have shown that up to a temperature of 400° C the coordination number is equal to 8. Consequently, upon melting the coordination number of sodium does not change.

Potassium. As in the case of sodium, it has a volume-centered lattice in the crystalline state. In liquid potassium up to a temperature of 390° C the coordination number is equal to 8, i.e., the volume-centered structure is retained.

Lithium. According to some insufficiently verified data the coordination number in liquid lithium at a temperature of 200° C is nearly 10, whereas in the crystalline state it is equal to 8, corresponding to a volume-centered lattice.

Bismuth. This belongs to those metals with the so-called porous grouping of the atoms in the solid state. Upon melting bismuth this grouping apparently is broken up, resulting in a coordination number increase from 6 in the crystal to 7-7.5 in the liquid.

Gallium. Just as in the case of bismuth, the porous grouping of atoms is broken up whengallium is melted, bringing it nearer a dense atomic grouping, in correspondence with which the coordination number increases from 6 to 11-12.

Lead. There are no reliable data on lead, but there is a basis for assuming that upon melting lead its coordination number does not change,but remains equal to 12.

The Main Physical Properties of Liquid Metals

In Tables 1.1-1.9 below the principal physical properties of certain liquid metals and their alloys are given. These data were obtained in the laboratory of Academician Mikheev by Nikol'skii, Kalakutskaia, Pchelkin, Klassen, and Vel'tishcheva.

The specific gravity of sodium and of sodium-potassium and lead-bismuth alloys was determined by the method of heavy suspension in the molten metal and by the pycnometer method. Values of the specific gravity of potassium, lithium, tin, lead, bismuth, and mercury are taken from several different published papers [22], [64-68].

TABLE 1.1
Mercury (t_{melt} = −38° C, t_{boil} = 357° C)

t °C	γ kg/m³	λ Cal/m·hr·° C	c_p Cal/kg·°C	$a\cdot10^2$ m²/hr	$\nu\cdot10^8$ m²/sec	$Pr\cdot10^2$
0	13 590	6,70	0,0334	1,48	12,4	3,02
10	13 570	6,81	0,0333	1,51	11,8	2,81
20	13 550	6,92	0,0332	1,54	11,4	2,66
50	13 470	7,25	0,0329	1,64	10,4	2,30
100	13 350	7,80	0,0328	1,78	9,4	1,90
150	13 230	8,35	0,0328	1,92	8,6	1,61
200	13 110	8,90	0,0328	2,07	8,0	1,39
250	13 000	9,45	0,0328	2,22	7,5	1,22
300	12 880	10,00	0,0328	2,37	7,1	1,08
350	12 800	10,50	0,0328	2,50	6,8	0,98
400	12 700	10,85	0,0329	2,60	6,6	0,91
450	12 600	11,15	0,0329	2,69	6,4	0,86
500	12 480	11,45	0,0330	2,78	6,2	0,80

TABLE 1.2
Sodium (t_{melt} = 97.3°C, t_{boil} = 878° C)

t °C	γ kg/m³	λ Cal/m·hr·°C	c_p Cal/kg·°C	$a\cdot10^2$ m²/hr	$\nu\cdot10^8$ m²/sec	$Pr\cdot10^2$
100	928	74,0	0,331	24,1	77,0	1,15
150	916	72,3	0,324	24,4	59,4	0,88
200	903	70,2	0,317	24,5	50,6	0,74
250	891	67,7	0,311	24,4	44,2	0,65
300	878	64,9	0,306	24,2	39,4	0,59
350	866	61,8	0,304	23,5	35,4	0,54
400	854	59,1	0,304	22,8	33,0	0,52
450	842	56,8	0,304	22,2	30,8	0,50
500	829	54,9	0,304	21,8	28,9	0,48
550	817	53,3	0,304	21,5	27,2	0,46
600	805	52,1	0,305	21,2	25,7	0,44
650	792	51,3	0,305	21,2	24,4	0,41
700	780	50,8	0,305	21,3	23,2	0,39

The coefficient of thermal conductivity of the metals was measured by the method of successive stationary states [77]. The coefficient of thermal conductivity of mercury was further measured over a wider temperature interval, using a column with compensation for lateral heat losses with height.

The specific heat of the metals was determined by the method of direct heating and by the method of comparison. For certain metals an extrapolation and averaging of the values of c_p were made.

The values of the kinematic viscosity of the metals (except for tin and mercury) were measured by the method of damped torsional oscillations, while for tin and mercury the values were taken from references [69-72].

TABLE 1.3
Potassium (t_{melt} = 63.7° C, t_{boil} = 760° C)

t °C	γ kg/m³	λ Cal/m·hr·°C	c_p Cal/kg·°C	$a \cdot 10^2$ m²/hr	$\nu \cdot 10^8$ m²/sec	$Pr \cdot 10^2$
100	818	40,0	0,195	25,1	56,1	0,80
150	807	39,9	0,192	25,8	48,0	0,67
200	795	39,5	0,189	26,3	42,8	0,59
250	784	38,6	0,187	26,3	38,6	0,53
300	773	37,3	0,185	26,1	35,2	0,49
350	761	35,8	0,184	25,6	32,4	0,46
400	750	34,0	0,183	24,8	29,8	0,43
450	738	32,0	0,183	23,7	27,6	0,42
500	727	30,0	0,183	22,6	25,7	0,41
550	716	28,2	0,184	21,4	24,1	0,41
600	704	26,6	0,184	20,5	22,1	0,39
650	692	25,3	0,185	19,8	21,6	0,39
700	681	24,3	0,185	19,3	20,5	0,38

TABLE 1.4
Lithium (t_{melt} = 186° C, t_{boil} = 1317° C)

t °C	γ kg/m³	λ Cal/m·hr·°C	c_p Cal/kg·°C	$a \cdot 10^2$ m²/hr	$\nu \cdot 10^8$ m²/sec	$Pr \cdot 10^2$
200	515	39,6	0,990	7,77	111,0	5,14
250	510	39,8	1,001	7,80	100,4	4,63
300	505	40,1	1,012	7,85	92,7	4,25
350	500	40,3	1,023	7,88	86,6	3,96
400	495	40,5	1,034	7,91	81,7	3,72
450	490	40,7	1,045	7,95	77,4	3,51
500	484	40,9	1,056	8,00	73,4	3,30
550	479	41,1	1,067	8,04	70,0	3,13
600	474	41,3	1,078	8,08	66,8	2,98
650	470	41,5	1,089	8,11	63,8	2,83
700	465	41,7	1,100	8,15	61,7	2,73

Data on the heat of fusion, the heat of vaporization, and the vapor tension of liquid metals are cited in [45]. Some information on the physical properties of gallium, cesium, and other metals in the melted state are given in [21] and [45].

TABLE 1.5
Bismuth (t_{melt} = 271° C, t_{boil} = 1490° C)

t °C	γ kg/m³	λ Cal/m·hr·°C	c_p Cal/kg·°C	$a \cdot 10^3$ m²/hr	$\nu \cdot 10^8$ m²/sec	$Pr \cdot 10^2$
280	10 050	12,5	0,036	3,45	18,0	1,88
300	10 030	12,6	0,036	3,50	17,1	1,76
350	9970	13,0	0,036	3,63	15,4	1,53
400	9910	13,4	0,036	3,76	14,2	1,36
450	9850	13,8	0,036	3,90	13,1	1,21
500	9785	14,2	0,036	4,03	12,2	1,09
550	9720	14,6	0,036	4,17	11,5	0,99
600	9660	14,9	0,036	4,31	10,8	0,91
650	9595	15,3	0,036	4,45	10,3	0,84
700	9530	15,7	0,036	4,59	9,8	0,77

TABLE 1.6
Tin (t_{melt} = 231.9° C, t_{boil} = 2270° C)

t °C	γ kg/m³	λ Cal/m·hr·°C	c_p Cal/kg·°C	$a \cdot 10^3$ m²/hr	$\nu \cdot 10^8$ m²/sec	$Pr \cdot 10^2$
240	6985	26,2	0,061	6,15	27,3	1,60
250	6980	26,4	0,061	6,20	26,7	1,55
300	6940	27,2	0,061	6,42	24,1	1,35
350	6905	28,0	0,061	6,65	21,9	1,19
400	6865	28,9	0,061	6,89	20,1	1,05
450	6830	29,7	0,061	7,13	18,6	0,94
500	6790	30,5	0,061	7,36	17,4	0,85
550	6755	31,4	0,061	7,62	16,5	0,78
600	6720	32,2	0,061	7,85	15,6	0,72
650	6680	33,0	0,061	8,10	15,0	0,67
700	6640	33,9	0,061	8,37	14,3	0,61

TABLE 1.7
Lead (t_{melt} = 327.4° C, t_{boil} = 1740° C)

t °C	γ kg/m³	λ Cal/m·hr·°C	c_p Cal/kg·°C	$a\cdot 10^2$ m²/hr	$\nu\cdot 10^8$ m²/sec	$Pr\cdot 10^2$
400	10 592	13,0	0,0352	3,49	21,0	2,17
450	10 536	13,2	0,0352	3,56	19,1	1,93
500	10 476	13,3	0,0352	3,61	17,5	1,75
550	10 419	13,4	0,0352	3,65	16,3	1,61
600	10 360	13,7	0,0352	3,76	15,3	1,47
650	10 300	14,4	0,0352	3,97	14,5	1,32
700	10 242	15,2	0,0352	4,22	13,7	1,17
750	10 185	16,2	0,0352	4,52	13,1	1,04
800	10 125	17,0	0,0352	4,77	12,5	0,94

TABLE 1.8
Sodium-Potassium Alloy (25% Na + 75% K)
(t_{melt} = − 11° C, t_{boil} = 784° C)

t °C	γ kg/m³	λ Cal/m·hr·°C	c_p Cal/kg·°C	$a\cdot 10^2$ m²/hr	$\nu\cdot 10^8$ m²/sec	$Pr\cdot 10^2$
20	872	19,0	0,310	7,03	93,0	4,76
50	865	19,5	0,294	7,67	78,5	3,69
100	852	20,0	0,273	8,60	60,7	2,51
150	840	20,5	0,260	9,39	51,7	1,98
200	828	21,1	0,256	9,95	45,2	1,64
250	815	21,6	0,252	10,52	40,3	1,38
300	803	22,2	0,248	11,15	36,6	1,18
350	790	22,7	0,244	11,78	33,4	1,02
400	778	23,3	0,240	12,48	30,8	0,89
450	765	23,8	0,235	13,24	28,2	0,77
500	753	24,4	0,231	14,03	26,7	0,69
550	741	24,9	0,227	14,80	25,2	0,61
600	729	25,5	0,223	15,69	23,7	0,54
650	716	26,0	0,219	16,58	22,5	0,49
700	704	26,6	0,215	17,57	21,4	0,44

TABLE 1.9

Lead-Bismuth Alloy (44.5% Pb + 55.5% Bi)

(t_{melt} = 123.5° C, t_{boil} = 1670° C)

t °C	γ kg/m³	λ Cal/m·hr·°C	c_p Cal/kg·°C	$a \cdot 10^2$ m²/hr	$\nu \cdot 10^8$ m²/sec	$Pr \cdot 10^2$
130	10 570	9,4	0,035	2,54	31,4	4,45
150	10 547	9,6	0,035	2,60	28,9	4,00
200	10 486	10,1	0,035	2,75	24,3	3,18
250	10 425	10,5	0,035	2,88	21,0	2,62
300	10 364	10,9	0,035	3,00	18,7	2,24
350	10 300	11,3	0,035	3,13	17,1	1,97
400	10 242	11,8	0,035	3,29	15,7	1,72
450	10 180	12,2	0,035	3,42	14,6	1,54
500	10 120	12,6	0,035	3,56	13,6	1,37
550	10 060	13,1	0,035	3,72	12,9	1,25
600	10 000	13,6	0,035	3,88	12,4	1,15
650	9 940	14,0	0,035	4,02	11,8	1,06
700	9 876	14,4	0,035	4,16	11,4	0,99

The Thermodynamic Similarity Principle of Real Substances

The significance of the similarity principle in the investigation of various problems in hydrodynamics and heat exchange is well known. Using this theory, it is possible to generalize the experimental data of one experiment and to extend the results obtained in this connection to all similar cases, i.e., to generalize the behavior of the considered phenomena.

The theory of similarity can be of no less importance in studying the physical properties of matter, in particular in studying the parameters of various heat-transfer media. On the basis of this theory it is possible to make quite satisfactory conclusions on the properties of interest for a given substance with respect to the known properties of its thermodynamic analog (or a substance which is similar to it thermodynamically), a fact which is of very significant interest from a practical point of view in comparing or seeking new heat-transfer media.

In other words, having studied experimentally the physical properties of one of the heat-transfer media, it is possible to use this data with sufficient accuracy in determining the very same properties of other heat-transfer media, without performing a new experiment, the only provision being that these heat-transfer media are thermodynamically similar to the original.

We shall explain briefly what the concept of thermodynamic similarity means and under what circumstances thermodynamically analogous substances may be singled out.

If, in addition to the thermal parameters — specific volume v, pressure p, and temperature of the body T — we introduce the new variables $\varphi = \frac{v}{v_K}$, $\pi = \frac{p}{p_K}$, $\tau = \frac{T}{T_K}$ (v_K, p_K, and T_K are the critical parameters of the given body), then the equation of state of the body can be written in dimensionless form:

$$\pi = f(\varphi, \tau, c). \tag{1.1}$$

This equation contains numerical coefficients which are identical for all substances and certain dimensionless constants c which depend on the nature of the substance. The equation of state is considered as a reduced equation of state, and the variables φ, π, τ are the reduced volume, pressure, and temperature, respectively.

States in which substances have identical reduced parameters are called corresponding states. Since at the critical point all three reduced parameters have the same value, which is equal to unity, it is clear that the critical states of all substances are corresponding ones.

If two or more substances satisfy one and the same reduced equation of state and have two of the three reduced parameters φ, π, τ, the same, then the third parameter will also be identical for these, i.e., the substances are in corresponding states. This is called the law of corresponding states.

Substances which obey this law and which satisfy the same reduced equation of state are called thermodynamically similar substances.

From the law of corresponding states, in particular, it follows that at the critical point the ratio $\frac{RT_K}{p_K v_K}$, which is called the critical coefficient, should have the same value for all thermodynamically similar substances.

For a number of thermodynamic quantities some very important consequences come out of the law of corresponding states. It is easily verified, for example, that the molar entropy S_M (or, more precisely, the difference in entropy $S_M - S_M^0$, where S_M^0 is the value of S_M when $T \to 0$), the ratios $\frac{U_M - U_M^0}{T}$ and $\frac{F_M - F_M^0}{T}$ (here U_M is the internal energy and F_M is the free energy of the body per mole) are functions of the reduced parameters and of the ratios c_{v0}/R or c_{p0}/R, where c_{v0} and c_{p0} are the values of the specific heats c_v and c_p at a given temperature and for $v \to \infty$ (or $p \to 0$); these functions are identical for all thermodynamically similar substances. As we know, with the help of the thermodynamic functions U, S, F, the processes of conversion of energy during a change of state of a body, e.g., during phase transitions, in the propagation and transfer of heat both from external sources and from the action of internal friction forces, in the increase or decrease of the surface of the body, etc., are described. Therefore, the law of corresponding states can be extended to different energy transformation processes, especially to the processes of heat propagation, phase transitions, etc. From this law, in particular, it follows that the specific heats c_v and c_p, the heat of vaporization $\underline{r}$, and the coefficient of surface tension σ, the viscosity μ, and the thermal conductivity λ in the liquid and gaseous states should be determined by the following general relations for all thermodynamically similar substances:

$$c_M = MRf_1\left(\pi,\ \tau,\ \frac{c_{M0}}{MR}\right), \tag{1.2}$$

$$Mr_{\text{boil}} = MRf_2(\tau), \tag{1.3}$$

$$\sigma = \sqrt[3]{MRT_K\, p_K^2}\, f_3\left(\tau,\ \frac{c_{M0}}{MR}\right), \tag{1.4}$$

$$\mu = \frac{\left(\frac{M}{g}\right)^{1/2} p_K^{2/3}}{(MR)^{1/6}} f_4\left(\pi,\ \tau,\ \frac{c_{M0}}{MR}\right), \tag{1.5}$$

$$\lambda = \frac{(MR)^{5/6} p_K^{2/3}}{\left(\frac{M}{g}\right)^{1/2} T_K^{1/6}} f_5\left(\pi,\ \tau,\ \frac{c_{M0}}{MR}\right). \tag{1.6}$$

In these equations the functions $f_1, \ldots, f_5$, are universal functions of the independent reduced parameters, M is the molecular weight of the substance, MR is the universal gas constant, and $\underline{g}$ is the gravitational acceleration.

The relations (1.2) - (1.6) for c, r, σ, μ, λ, make it possible from the known properties of one substance to determine the properties of other thermodynamically similar substances under the same conditions, i.e., for equal values of the reduced parameters.

The indicated general expressions for c, r, σ, μ, λ and the analogous relations for the other physical properties can be derived on the basis of the law of corresponding states in a number of ways. We shall cite one of these.

The dependence of any of the quantities which characterize the physical properties of a body on the state parameters can be represented in the form of the product of a factor having the same dimensionality as the investigated substance and composed of the principal physical parameters of the given body — p_K, T_K, M, MR — by a dimensionless function of the reduced parameters π, τ (or φ, τ) and the ratio c_{M0}/MR. Combining these parameters and their powers in such a way that the dimensionality of the quantity of interest is obtained, it is not difficult to obtain an expression for a dimensional multiplier of any of the desired relations.

It is well known, however, that the law of corresponding states does not have a completely general applicability, but is valid only within certain definite groups of substances. In connection with this, the general relations for the physical properties also are applied only for groups of thermodynamically similar substances.

The question naturally arises as to the means for classifying substances according to these groups, i.e., the criteria for thermodynamic similarity.

As shown by an analysis of this problem, the shape of the potential curve for the van der Waals interaction of the molecules of the given substance is of considerable importance. The reason for this becomes clear if we take into account the fact that in the equation of state there appear only the individual (depending on the nature of the given substance) constants which are contained in the analytic expression for the potential energy of the van der Waals interaction between two molecules u(r) as a function of the distance r between them. If the number of these individual constants were not to exceed two, then they could be excluded (with the help of the two conditions determining the critical point) from the equation of state, which could be reduced to a dimensionless expression with no constants depending on the nature of the substance. In this case, the law of corresponding states would be a general law, i.e., it would be valid for all substances. In actuality, the number of individual constants appearing in the expression for the potential energy of the van der Waals interaction is greater than two. Therefore, a single reduced equation of state which is general for all substances without exception does not exist and the law of corresponding states does not have a comprehensive applicability, only a limited one, i.e., it is valid only for thermodynamically similar substances.

TABLE 1.10
Thermodynamic Characteristics of Specific Substances

Substance	$-\frac{u_0}{kT_{\text{к}}}$	$\frac{RT_{\text{к}}}{p_{\text{к}} v_{\text{к}}}$
He	~1	3,27
H_2	0,935	3,28
N_2	0,767	3,43
A	0,766	3,43
O_2	0,695	3,43
CH_4	0,657	3,45

On the basis of the theory of the equation of state of real substances it is possible to establish the following general conditions or criteria of thermodynamic similarity:

1. Two or more substances are thermodynamically similar if the values of the potential energy of the van der Waals interaction of two molecules are proportional, i.e., if $u_1(r) = \text{const}\, u_2(r)$ (here r is the distance between centers of the two molecules). Geometrically, this condition implies a similarity in the curves of the potential energy of the van der Waals interaction for the considered substances.

2. Two or more different substances are thermodynamically similar if the following dependence for the potential energy of the van der Waals molecular interaction exists:

$$u_1(r_1) = \text{const}\, u_2(r_2) \quad \text{for} \quad \frac{r_1}{r_2} = \text{const}.$$

The first condition is a special case of this relation. This general dependence was established in a somewhat different form from the sufficiently rigorous considerations of Kammerling and Onnes.

The indicated conditions have the same general significance and are valid for any functional dependence u(r), i.e., for any number of individual constants appearing in the expression for u. If we consider that in u(r) in the first approximation there appear three individual constants which characterize the nature of the substance, then it is possible to establish a more specific, but concurrently more practical third criterion for thermodynamic similarity, which can be formulated as follows:

3. Two or more different substances are thermodynamically similar if the ratio of the equilibrium energy u_0 (i.e., the value of the potential energy of the van der Waals molecular interation at the minimum point) to the critical temperature is the same for all of them:

$$-\frac{u_0}{kT_{\text{к}}} = \text{const}.$$

The above discussion is well illustrated in Table 1.10, which contains the values of the ratio $-u_0/kT_K$ and the critical coefficient $RT_K/p_K v_K$.

There are still several conditions, but these are analogous to the third condition. Essentially they all refer to the case in which for some substances one of the three individual constants either has an equal value or between two of the constants there is a numerical relation which is general for all these substances. Then one of the three constants can be regarded as a constant of the group rather than an individual one and is general for the entire group of given substances. The number of individual constants for each of the substances in a given group diminishes to two and, consequently, within this group the law of corresponding states is valid.

Application of the Thermodynamic Similarity Principle to the Study of The Properties of Liquid Metals

Within the framework of each group of thermodynamically similar substances the physical properties of a substance (viscosity, thermal conductivity, surface tension, etc.) can be expressed as a functional identity in the form of a product of a dimensional factor consisting of definite dimensional quantities (molecular weight and the critical parameters) by a function of the reduced parameters which is the same for all substances of a given group. In particular, the dependence of the viscosity coefficient on the state parameters has the form

$$\mu = \left(\frac{M}{g}\right)^{1/2} \frac{p_K^{2/3}}{(MR)^{1/6}} f\left(\pi, \tau, \frac{c_{M^0}}{MR}\right). \tag{1.7}$$

In the liquid state at temperatures considerably less than T_K the dependence of the viscosity on pressure is extremely insignificant, while the effect of c_{M0} is small, so that

$$\mu = \left(\frac{M}{g}\right)^{1/2} \frac{p_K^{2/3}}{(MR)^{1/6}} f(\tau). \tag{1.8}$$

The viscosity of liquid metals has been investigated by a number of authors. The viscosity of liquid metals was studied rather comprehensively by Solov'ev [22], [57]. As a result of these investigations, it was established that the various liquid metals could be split into several groups on the basis of the criterion for thermodynamic similarity. For example, sodium, potassium, rubidium, and cesium (elements in the left-hand subgroup 1 of the Mendeleev periodic table), which crystallize into a cube-centered lattice and have approximately equal values for the coordination number, compose one group, while gallium, bismuth, antimony, mercury, and tin (elements in the right-hand subgroup 1 of the periodic table) compose a second group; lithium, however, should be referred to this group since it has a coordination number which is almost equal to 10 in the liquid state.

For reassurance of this fact, let us consider Fig. 1.1. The values of the logarithm of the ratio of the coefficient of viscosity μ at a given temperature to the viscosity coefficient μ_m at the melting point are plotted vertically, and the values of the ratio T/T_m are plotted horizontally. The melting temperature T_m can with a sufficient degree of accuracy be considered as nearly corresponding to the critical temperature, so that instead of the ratio T/T_K we can choose the ratio T/T_m. Then from the general expression for μ we find

$$\frac{\mu}{\mu_m} = \phi\left(\frac{T}{T_m}\right).$$

From this equation it follows that the points which define the temperature dependence of μ on T in the μ/μ_m (or $\ln \mu/\mu_m$), T/T_m coordinate system should lie on one curve for thermodynamically similar substances. From the figure it is evident that this is actually the case: all the points for sodium, potassium, rubidium, and cesium are distributed along curve I, while the points for tin, mercury, antimony, bismuth, gallium, and lithium lie along curve II.

Thus, the experimental data lend considerable support to the possible separation of liquid metals into groups of thermodynamically similar substances.

The cited graphical dependence of μ on $\underline{t}$ also illustrates quite well the role of the thermodynamic similarity principle in studying the physical properties of substances. Having evolved tests, for example, with liquid sodium and having constructed a curve for the dependence of the viscosity of sodium on temperature in reduced coordinates, from this curve it is possible without performing a new experiment to calculate the values of the viscosity of sodium or rubidium under similar conditions.

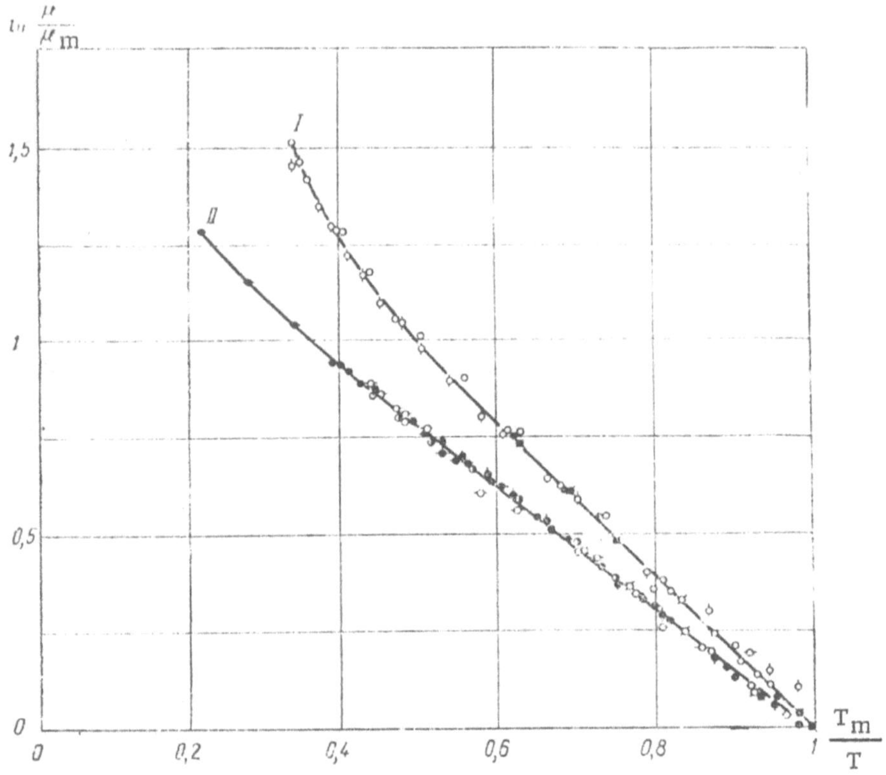

Fig. 1.1. Experimental data for the viscosity of various liquid metals.
I: ○) sodium, ♁) potassium, ✕) rubidium, ⚲) cesium;
II: ●) tin, ◑) mercury, ✕) antimony, ⧫) bismuth, –●–) gallium
–○–) lithium.

Experimental Data on the Sound Propagation Velocity in Liquid Metals

Table 1.11 gives data on the velocity of sound in a number of metallic liquids near the melting point, as obtained by Kleppo [52].

As we know, the velocity of sound in water at t = 20° C is equal to 1410 m/sec. From this, it is evident that in liquid metals the value of $\underline{a}$ is of the same order of magnitude as it is in nonmetallic liquids.

Computational Estimation of the Velocity of Sound in Liquid Metals

A semiempirical method for determining the velocity of sound was proposed by Krzhizhanovskii. If the adiabatic compressibility β_{ad} is known, the velocity of sound can be determined by the equation

$$a = \sqrt{\frac{1}{\beta_{ad}\rho'}} \,. \tag{1.9}$$

According to Jacobson's theory [53] the quantity β_{ad} is proportional to the molecular range Λ, where the proportionality factor depends only on the temperature. Hence,

$$a = \frac{k}{\Lambda \sqrt{\rho'}}\,. \qquad (1.10)$$

The quantity Λ is determined by the equation

$$\Lambda = \frac{2\,(V - V_0)}{\sqrt[3]{36\pi N V_0^2}}\,, \qquad (1.11)$$

where N is Avogadro's number, V_0 (m^3/mole) is the volume of the liquid when the atoms are densely grouped.

TABLE 1.11

Sound Propagation Velocity in Liquid Metals Near the Melting Point

Substance	Testing temperature, °C	Velocity of sound a m/sec	Temperature coefficient $\frac{\partial a}{\partial t}$ m/deg · sec
Zn	420	2790±60	—
Ga	30	2740	—
Cd	321	2200	—0,5
In	156	2215	—0,5
Sn	232	2270	—0,7
Te	302	1625	—
Pb	327	1790	—0,5
Bi	271	1635	—0,5
Na	98	2395	—0,3
K	64	1820	—0,5
Rb	39	1260	—0,4
Cz	29	967	—0,3
Eutectic			
Sn—Cd	250	2205	—0,6
Sn—Hg	200	1950	—0,3
Sn—Bi	200	1965	—
Sn—Pb	290	1975	—

TABLE 1.12

Values of the Coefficient k [Equation (1.10)] Obtained in Experiments With Nonmetallic Liquids

t, °C	0	10	20	25	30	40	50
$k \cdot 10^8$, kg $^{1/2}$	588	604	618	625	631	642	652

Substituting the value for Λ into the expression (1.10), we finally obtain

$$a = \frac{k}{2\,(1-\xi)\sqrt{\rho'}} \sqrt[3]{\frac{36\pi N \xi^2}{V}}\,, \qquad (1.12)$$

where $\xi = \frac{V}{V_0}$.

In Fig. 1.2 the values of k are shown extrapolated to a temperature of 140° C.

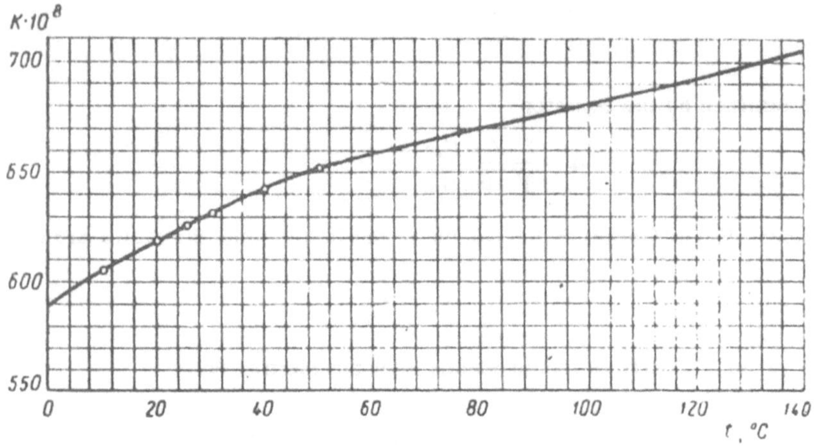

Fig. 1.2. Temperature dependence of the coefficient k.

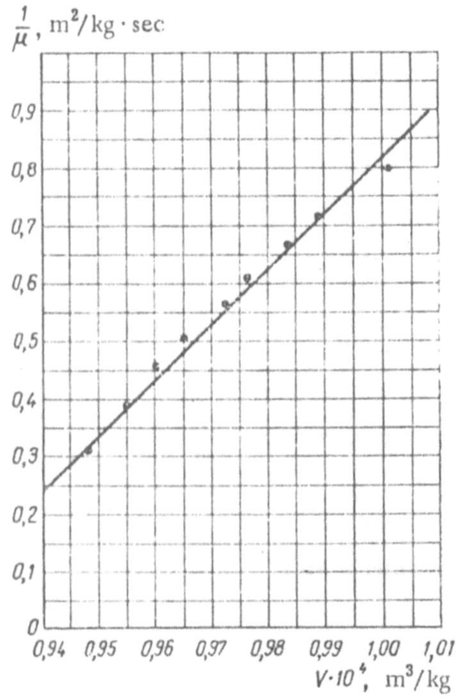

Fig. 1.3. Viscosity coefficient as a function of specific volume for a eutectic lead-bismuth alloy.

The quantity $\Delta V = V - V_0$ for many nonmetallic liquids is quite closely connected with the dynamic viscosity coefficient of Bachinskii's equation

$$\mu \Delta V = \text{const}, \tag{1.13}$$

or

$$\frac{1}{\mu} = \frac{1}{\text{const}} (V - V_0). \tag{1.14}$$

Figure 1.3 shows the function $\frac{1}{\mu} = f(V)$ for a eutectic lead-bismuth alloy, constructed from experimental data. To a first approximation the linear behavior of this function is preserved for any metallic liquid.

Consequently, in this case, when there are no directly determined values of a or β_{ad}, the velocity of sound can be calculated by means of (1.12) and data on the viscosity coefficient and specific volume. The accuracy of computation is limited by the necessity of extrapolating the values of the coefficient k in the high temperature region and by the necessity of observing the relation (1.13) for a given metallic liquid.

Chapter 2

AREAS OF APPLICATION FOR LIQUID-METAL HEAT-TRANSFER MEDIA

General Discussion

The main advantages of liquid-metal heat-exchangers in comparison with other heat-exchangers are contained in the following:

1. Liquid metals are characterized by exceptionally large values of the coefficient of heat conductivity. Given an outflow of the same quantity of heat, as a result of the comparatively large specific heat of liquid metals the amount of the heat-transfer media required is less in comparison with other types. In this case the behavior of flow in liquid metals is similar to the behavior of ordinary liquids, as a result of which the pressure losses in changing over a heat-withdrawal system are not excessive (see Chapter 3).

2. Owing to the high boiling temperature, the process of heat transfer and outflow under conditions of higher operating temperatures can be realized at comparatively low pressures. The magnitude of the pressure of saturated liquid-metal vapors has the advantage that cavitation in the liquid-metal flow does not develop to any great extent.

3. As a consequence of the comparatively low melting temperature (for sodium t_m is equal to 97.5° C, while for eutectic sodium-potassium it is 11° C), these metals are easily converted to the liquid phase and in such form are kept in the apparatus.

4. Liquid (in particular, alkali) metals give rise to comparatively weak corrosion of construction materials, as a result of which for the latter it is possible to use steel extensively, along with other metals and their alloys.

The shortcoming of some liquid metals, for example, sodium or potassium, is their violent chemical interaction with water and oxygen, which requires special precautionary measures in their handling.

Application of Liquid-Metal Heat-Exchangers in Steam Power Plants

It is a familiar fact that the comparatively low critical temperature of water limits the possibility of increasing the temperature of the vapor used in a turbine and thus impedes any substantial increase in the efficiency of the thermal cycle. For example, in order to have a mean temperature of the operating substance in the process of delivering heat to the cycle equal to 550° C, it is necessary to have a water pressure in the apparatus of more than 300 atmos. From this fact, it is evident that even with a very high pressure, when the steam power plant becomes extremely bulky and costly, the application of water vapor as an operating substance does not permit the most satisfactory means for utilizing the temperature limit which has already been accomplished, owing to the achievements of modern metallurgy.

Liquid-metal heat-exchangers better satisfy the thermodynamic requirement which the operating substance is called upon to give in steam power plants. The critical temperature of the operating substance should be considerably higher than the extreme temperatures applied at the present time in these plants.

For mercury, as an example, the critical temperature T_c is equal to 1400° C, i.e., four times greater than in the case of water. However, liquid-metal heat-exchangers have a gross disadvantage which limits their application in steam power plants. Essentially, this lies in the fact that the operating substance of a steam power plant at the lowest temperatures in the cycle, which usually amount to 25-28° C, should not have an excessively small vapor-saturation pressure, since in the opposite case the maintenance of a very high vacuum in the condenser would be extremely difficult.

Since the application of liquid-metal heat-transfer media, owing to their high critical temperature, was very enticing and promised a marked increase in the efficiency of operation, there arose the idea of applying two operating substances in the steam power plant over a definite temperature interval, relative to which the vapor-saturation pressure of a given substance would be most applicable.

Steam power plants with two operating substances have been given the name of binary heat engines. So far, in practice only mercury-water engines have been realized [3]. A schematic diagram of such a binary heat engine is shown in Fig. 2.1.

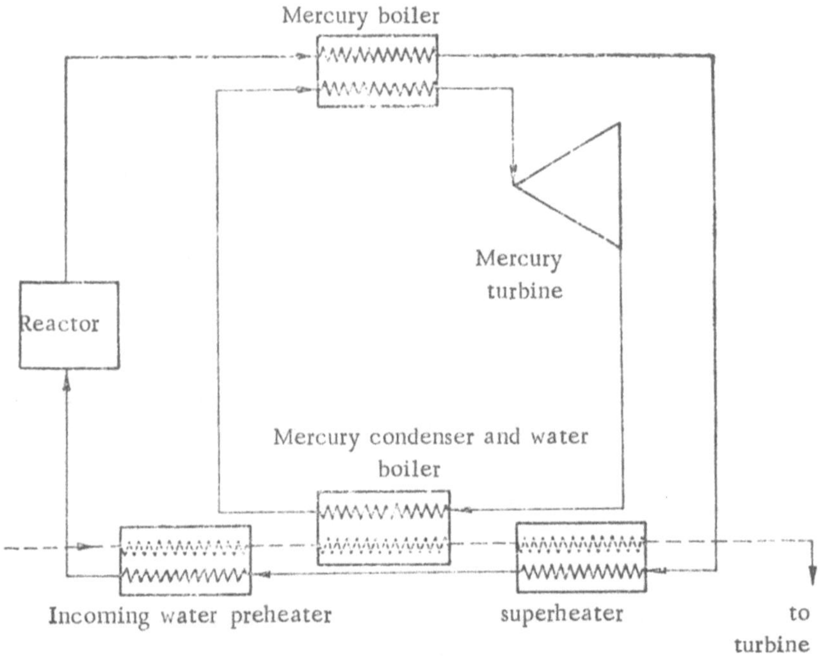

Fig. 2.1. Schematic diagram of a binary (mercury-water) heat engine.

Mercury vapor, which is formed in a mercury boiler, emerges from the boiler into a turbine. After expansion in the turbine the mercury vapor is directed into a so-called condenser-vaporizer, where it is condensed, during which the heat liberated in condensation is used for vaporizing the water. The liquid mercury from the condenser-vaporizer is once again directed into the mercury boiler, while the water vapor is passed into the steam superheater, after which it is delivered to a steam turbine.

In binary (mercury-water) heat engines the thermal efficiency of cycle amounts to 57% and higher, i.e., it exceeds the thermal efficiency of ordinary steam power plants operating at an average pressure [3].

Application of Liquid-Metal Heat-Exchangers in Nuclear Power Plants

The heat developing elements of modern nuclear reactors are essential for maintaining sufficiently high temperatures (up to 500-650° C) and thus they make it possible to carry off heat from the reactor at a comparatively high temperature level, which exerts a substantial influence on the efficiency of the nuclear power plant. Under these conditions the application of water as a heat-transfer medium becomes quite difficult, since the support of a high temperature of the heat-transfer medium in the reactor implies the necessity of increasing the water pressure to very large values. Therefore, in high-temperature reactors it is convenient for the removal of heat to apply liquid metals, which can be cycled under high temperature conditions at comparatively low pressures; moreover, some of these heat-transfer medium (for example, sodium) possess a smaller neutron-capture cross section in comparison with water, which is a circumstance of no small significance. However, liquid metals as heat-transfer media for reactors again have specific disadvantages. Being subjected to neutron radiation in the reactor, they become radioactive, which poses serious difficulties in the construction and servicing of the heat-withdrawal system of the reactor.

Chapter 3

HYDRAULIC RESISTANCE IN THE FLOW OF LIQUID METALS

Flow in Smooth Pipes

The essential feature of many molten metals as opposed to other liquid heat-transfer media is the fact that they do not wet solid surfaces around which they are flowing. For example, mercury does not wet carbon steel at a temperature less than 600° C, lead-bismuth at ordinary temperatures (250-600° C) does not wet carbon steel and tempered steel with a nickel content of less than 20-25%.

The idea has been expressed (with support in the single experiments of Poiseuille in 1843 and Tamman and Ginzberg in 1927) concerning the possibility of the liquids slipping without wetting the walls of a tube. In this case the velocity of the liquid at the walls would not be equal to zero and the laws of the hydraulic resistance (drag) of liquids which wet and do not wet the contact surface would not be identical.

Of the four experiments conducted by Barbour in 1870 on mercury flowing out from a glass capillary, two of the experiments obeyed the Poiseuille equation for laminar flow of a liquid in a circular tube, and two resulted in a marked deviation.

The experiments conducted by Tortman in 1928 with water ($Re \leq 5 \cdot 10^4$) and mercury ($Re \leq 5 \cdot 10^4$) flowing in steel pipes with a diameter of 1.2-15.4 mm gave concurrent results.

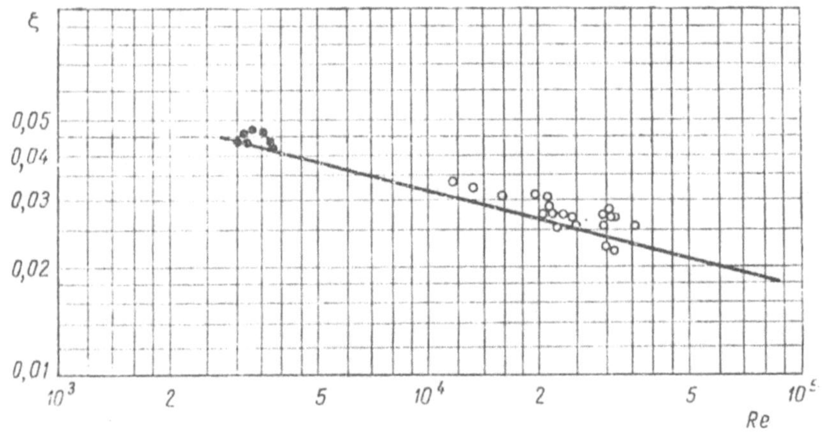

Fig. 3.1. Experiments with mercury and water in a glass tube.
——) represents the smooth tube equation; ●)—water, ○)—mercury.

Similar experimental investigations of the hydraulic resistance of mercury flowing in tubes were conducted in the USSR, beginning in 1934. Figure 3.1 shows the results of a determination of the resistance coefficient for mercury flowing in a smooth glass tube, as obtained by Syrkin. The experiments were conducted in the region of fully developed turbulent flow, and the results do not disclose any marked deviation from the curve corresponding to the usual Blasius dependence for these values of the Re number (from $5 \cdot 10^3$ to $100 \cdot 10^3$):

$$\zeta = \frac{0.316}{Re^{0.25}} . \tag{3.1}$$

Figure 3.2 shows the experimental data for mercury (obtained by Kondrat'ev [75]) and tin (obtained by Fedynskii and Osipov [75]) flowing in smooth steel pipes, and also the experimental data of Rightmire and Taylor [13], obtained in the flow of molten iron and pig iron in a ceramic tube.

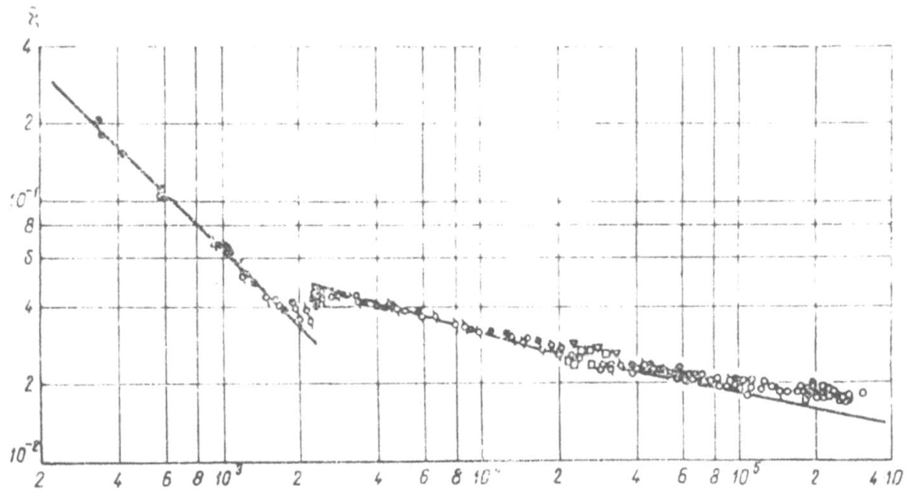

Fig. 3.2. Juxtaposition of the experimental data obtained by different authors for tin and mercury. Smooth steel tubes: ———) represents the smooth tube equation; ○) tin (D = 10.5 mm), ⊘) mercury (D = 4 and 5 mm), ●) water (D = 4, 5, and 10.5 mm). Ceramic tubes: □) molten iron, ▽) molten pig iron.

All of the experimental points are distributed around the analytic curve, where the identical behavior of the dependence of ζ on Re is distinctly visible.

Measurements carried out by Styrikovich, Sorin, and Semenovker [1, 2] showed that the velocity profile in the flow of mercury in a circular tube is the same as in the flow of nonmetallic fluids. The later experiments of Isakoff and Drew [17] followed this result (Fig. 3.3).

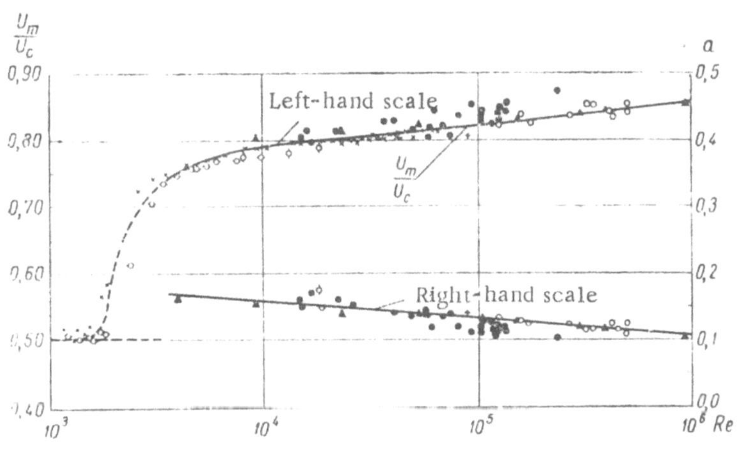

Fig. 3.3. Velocity profile associated with flow in a tube. ○)–Mercury (Isakoff and Drew), ●)–water (Cope), Δ)–water (Nikuradze), +) air (Stanton), ×)–water (Stanton and Pannell), -○-) air (Stanton and Pannell), ⦶)–water and air (Feidge); u_m is the mean velocity, and u_c is the maximum velocity.

Therefore, we can consider that the law of resistance for liquid metal flow in smooth tubes corresponds to the law for nonmetallic liquids with the accuracy necessary for the usual hydraulic calculations.

Flow in Rough Tubes

It is a well-known fact that in the flow of liquids in rough tubes, after attaining a definite value for the Re number, the roughness does not affect the hydraulic resistance coefficient. The effective quantity is the so-called square law of resistance, in which the quantity ζ depends only on the type of roughness and the relative dimensions of the roughness protuberances. In this case, the influence of wetting is absent in connection with the fact that the mechanism of the hydraulic resistance is not connected with the friction of the liquid against the over-all roughness of the tube, but with the formation of vortices in flowing around the individual protuberances.

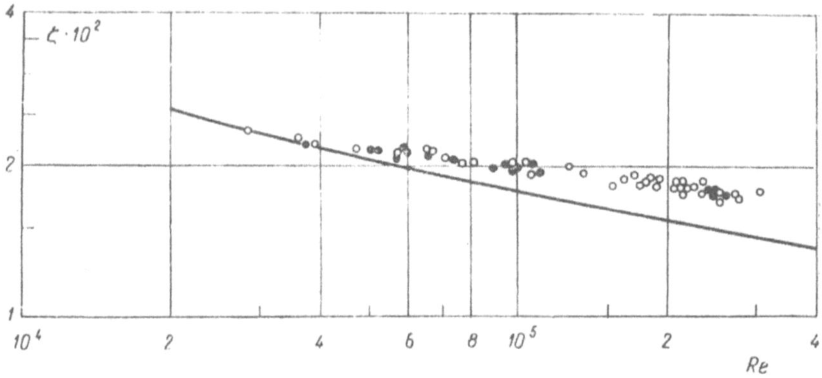

Fig. 3.4. Hydraulic resistance coefficient in the flow of molten tin in a steel tube having industrial roughness (D = 10.5 mm).
○)—Burnished internal roughness, ●)—plated internal roughness.

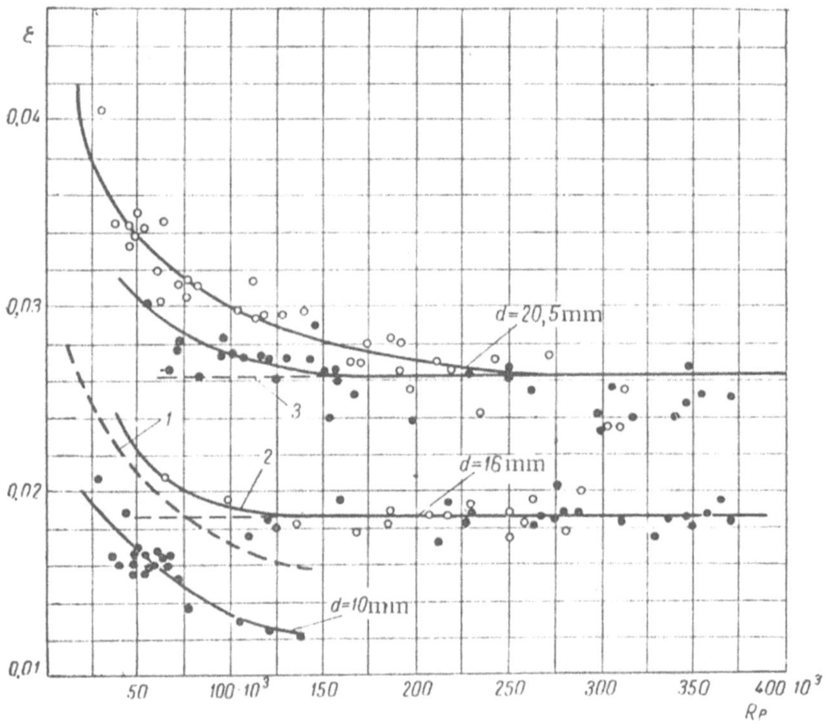

Fig. 3.5. Hydraulic resistance in the flow of mercury in steel tubes.
○)—Water, ●)—mercury, ------) analytic curves; 1) Blasius' data, 2) Nikuradze's data, with $\epsilon = 0.015$, 3) with $\epsilon = 0.06$.

In the experiments of Fedynskii and Osipov the hydraulic resistance coefficient was determined in the turbulent flow of molten tin in steel tubes having industrial roughness, with acidified (burnished) and plated internal roughnesses (Fig. 3.4).

It is evident that the results are practically identical both for wetted and for nonwetted surfaces.

The isothermal flow of mercury in rough steel pipes having a diameter of 19 and 32 mm at temperatures of 400-500° C and Re numbers from $1 \cdot 10^5$ to $3.5 \cdot 10^5$ have been investigated by Styrikovich, Sorin, and Semenovker [1, 2]. A value ζ = const was obtained, corresponding to the flow of nonmetallic liquids. A number of experiments on the flow of mercury in industrial steel pipes were conducted by Lozhkin, Kanaev, Gel'man [3, 4] et al. Figure 3.5 shows some of the results obtained by these authors. We see here the even behavior, characteristic for nonuniform roughness, of the curves in the interval where the square law of resistance is operative.

Figure 3.6 shows the results of the experiments of Borishanskii and Kutateladze [5, 33] in a study of the flow of bismuth in seamless pipe having a diameter of approximately 9 mm and made from stainless steel. The experiments were conducted with bismuth temperatures of 360-470° C. The results of these experiments agree with data obtained in the investigation of mercury.

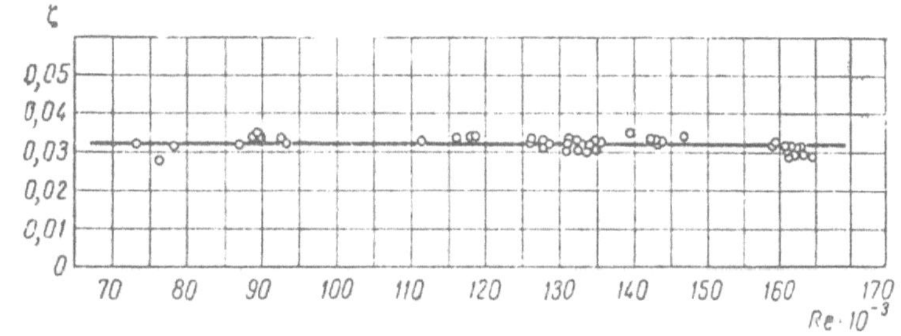

Fig. 3.6. Hydraulic resistance coefficient for bismuth flowing in a tube (D ≈ 9 mm).

Effect of Heat Flow

The high termal conductivity and comparatively weak dependence of the thermal-physical characteristics of liquid metals on temperature lead to the fact that the temperature profile in a liquid flow depends comparatively weakly on the magnitude of the heat flow. As a consequence of this the heat flow should have a slight effect on the hydraulic resistance in flowing liquid metals.

In the interval where the square law of resistance is operative this effect is particularly small, since in this case the hydraulic resistance depends only on one physical characteristic of the medium, namely, the density, which for liquid metals varies little with temperature.

An experimental verification of the effect of heat flow on the hydraulic resistance of a liquid metal was carried out by Ivashchenko. As a working substance eutectic lead-bismuth was chosen. The relatively large thermal conductivity of this alloy insured a significant change in temperature over the cross section of the liquid flow, even with moderate thermal loads.

In an experimental tube having a diameter of 10 mm with $L/D = 52$ tests were conducted with the isothermal flow of water, the isothermal flow of the alloy at $t_0 = 200$-280° C, and tests with the nonisothermal alloy at $t_0 = 170$-280° C and the numbers $Re = (40$-$150) \cdot 10^3$. In the last case the heat flow varied stepwise from 10^5 to $8 \cdot 10^5$ kcal/m^2·hour.

The results of the experiments are presented in graphical form in Fig. 3.7. As we can see, within the limits of experimental accuracy the effect of heat flow on the hydraulic resistance coefficient is not observable. The physical characteristics in the calculations are referred to the flow temperature t_0, averaged over the heat content.

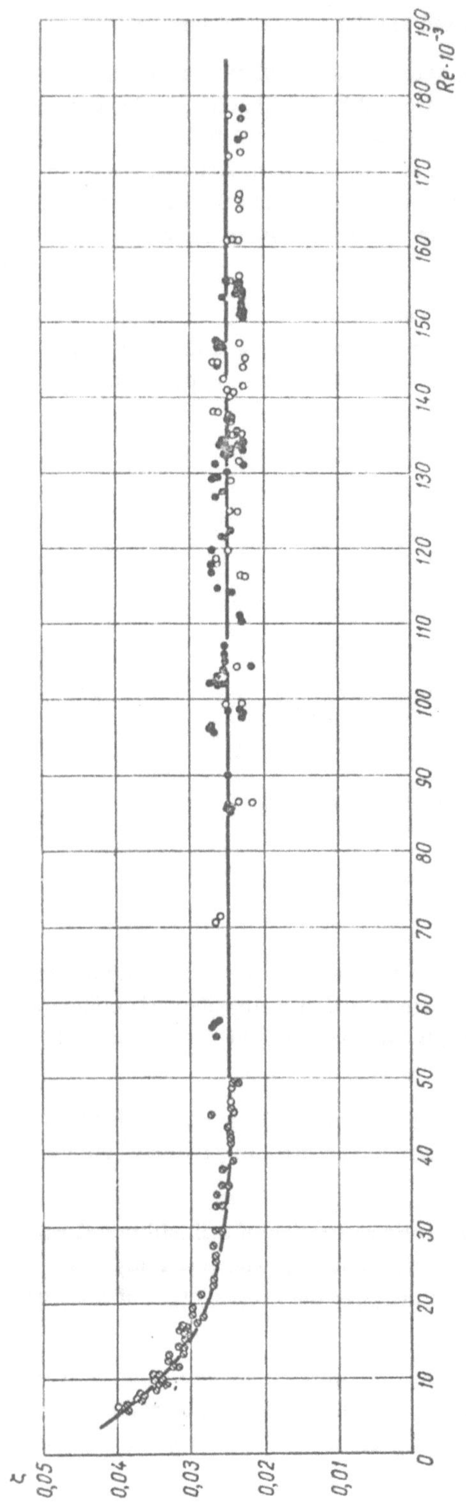

Fig. 3.7. Hydraulic resistance coefficient in the nonisothermal flow of a eutectic lead-bismuth melt in a steel tube, L / D = 52. Isothermal flow: ⊘)—Water, O)—melt. Nonisothermal flow: ●)—melt; $0 < q < 8 \cdot 10^5$ kcal/m²·hour.

Local Resistance

Sources of local hydraulic resistance: sudden variations in the cross section of a tubular duct, turns, shut-off and regulating mechanisms, diaphragms, grids, etc., are all characterized by the fact that in these the pressure drop due to friction is very small in comparison with the pressure drop caused by complete reordering of the velocity profile.

Thus, problems which are connected with liquid in contact with a wall are principally insignificant in the present case. In order to verify this assumption Kutateladze and Shneiderman [5] conducted tests on the flow of mercury and water through openings of various shapes.

The results of two series of these experiments are presented in Fig. 3.8. The square root of the initial volume of the liquid contained in a reservoir tank is plotted on the vertical axis, while the time required for the tank to be emptied during the outflow of the liquid through a given opening is plotted on the horizontal. As evident from this, the points for both liquids are distributed along single straight lines.

Repeated calibrations of test washers in water, mercury, tin, and a number of other molten metals also failed to detect any difference in the local hydraulic resistance coefficients in the flow of water and that of a molten metal.

Friction of a Rotating Disk

Figure 3.9 shows the results of experiments by Kanaev and Gel'man [3, 4] in determining the power expended in the rotation of a smooth disk having a diameter of 270 mm and thickness of 10 mm in various liquids.

In Fig. 3.10, these data are compared with the data of Gibson, Ryden, and Schmidt [3] and are presented so as to show the coefficient of friction as a function ζ of the number $Re = \frac{R^2 \omega}{\nu}$ (R is the radius of the disk and ω is the angular velocity of the disk).

The frictional moment of both surfaces of the disk was determined from the following equation ([3, 4]):

$$M_f = \zeta \frac{\gamma \omega}{2g} R^5. \qquad (3.2)$$

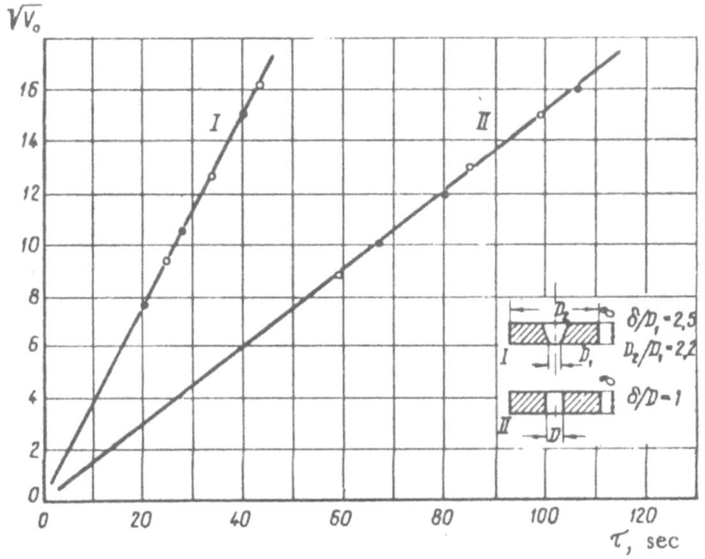

Fig. 3.8. Outflow time of mercury and water through a conical and a cylindrical nozzle (V_0 is the initial volume of the liquid). ●)—Mercury; ○)—water.

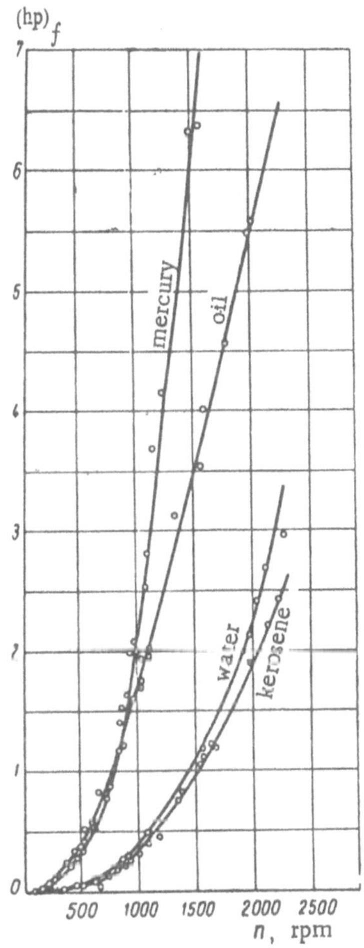

Fig. 3.9. Power expended in the rotation of a smooth disk in various liquids.

Increase in Pressure During Hydraulic Shock

The rapid closing of a shut-off mechanism gives rise to a sharp increase in pressure in a tubular conduit (so-called hydraulic shock), and this can destroy the conduit and its accessories.

Direct hydraulic shock arises whenever the shut-off time is less than twice the time required for a shock wave to pass through the conduit:

$$\tau < \frac{2L}{w^*}, \tag{3.3}$$

where τ is the shut-off time in seconds, w^* is the shock wave propagation velocity in m/sec, and L is the length of the conduit in m.

The magnitude of the shock wave velocity in the conduit is determined by Zhukovskii's equation [51]:

$$w^* = \frac{a}{\sqrt{1 + \frac{D}{\delta}\frac{E'}{E}}}, \tag{3.4}$$

where $\underline{a}$ is the sound propagation velocity of the liquid in the conduit in m/sec, D is the inside diameter of the conduit in m, δ is the thickness of the conduit wall in m, E is the elastic modulus of the material of the conduit in kg/m^2, and E' is the elastic modulus of the liquid in kg/m^2.

The elastic modulus of the liquid and the velocity of sound are related by the equation

$$E' = \rho' a^2, \tag{3.5}$$

where ρ' is the density of the liquid.

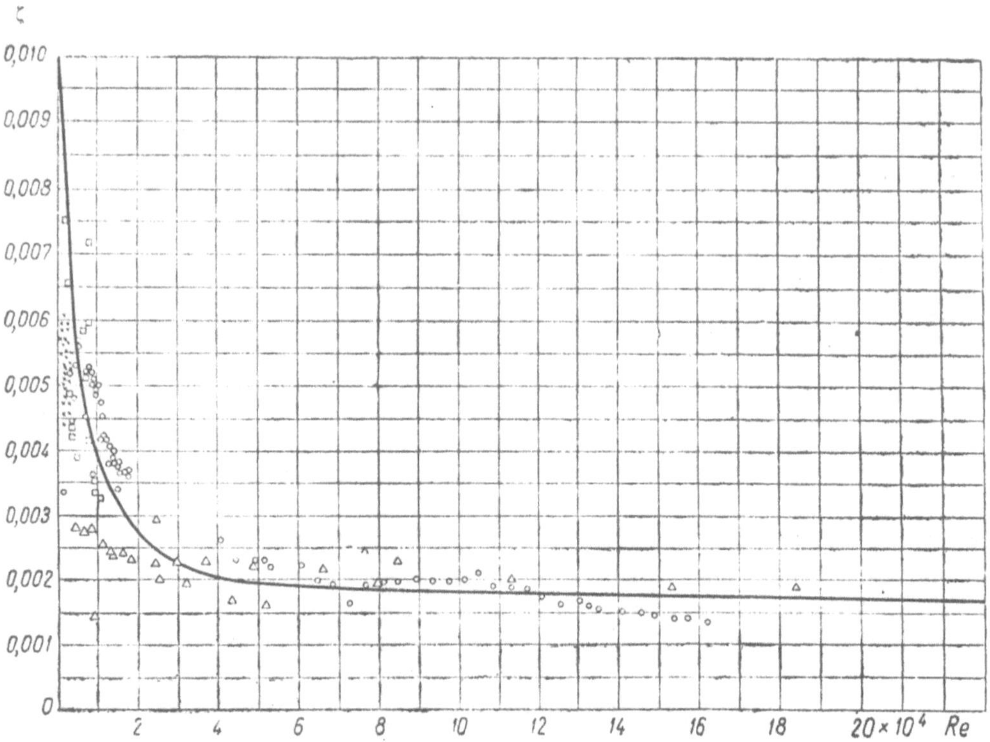

Fig. 3.10. Coefficient of friction of the disk as a function of the Re number.
O)–mercury, □)–kerosene, ⊖)–water: D = 270 mm (Kanaev, Gel'man, and Lozhkin);
Δ)–water, D = 200–500 mm (Gibson, Ryden); x)–water, D = 59–118 mm (Schmidt).

The pressure increase arising in the conduit during direct hydraulic shock is determined from the equation

$$\Delta p_{sh} = \rho' (w_0 - w) w^*, \tag{3.6}$$

where w_0 and $\underline{w}$ are the flow velocities of the liquid before and after shut-off (it is possible for the shut-off mechanism not to be completely closed, in which case $w > 0$).

Substituting the value for w^* into (3.6) and maintaining due regard for (3.5), we obtain

$$\Delta p = \frac{\rho' (w_0 - w) a}{\sqrt{1 + \frac{D}{\delta} \frac{\rho' a^2}{E}}}. \tag{3.7}$$

Therefore, for the calculation of the pressure increase during hydraulic shock it is necessary to know the conditions of flow (w_0 and $\underline{w}$), the dimensions of the conduit (D and δ), the elastic modulus of the conduit material and physical properties such as the density of the liquid and the velocity of sound in the liquid.

Chapter 4

TURBULENT HEAT TRANSFER IN LIQUID METALS

Classification of Heat-Transfer Media According To Their Kinematic Viscosity Coefficients and Diffusivities

All fluids (liquid and gaseous), depending on the influence of their properties on the conditions of heat transfer, can be grouped into three basic classes, which are characterized by the order of their Prandtl numbers:

$$Pr = \frac{\nu}{a} = \frac{g\mu c_p}{\lambda}.$$

Fluids for which $Pr \approx 1$ (gases at any temperatures and certain liquids in a narrow temperature range). For this class of fluids the ratio of the rates of molecular transfer of heat and momentum is of the order of unity, i.e., both effects extend to the same depth over the flow cross section, while the turbulent transfer plays an effective role only in the so-called turbulent nucleus of the flow. The thermal and dynamic boundary layers have the same thickness (for $Pr = 1$ and grad $p = 0$, δ_T is exactly equal to δ).

Fluids for which $Pr > 1$ (all nonmetallic liquids). In this class of fluids the molecular transfer of momentum is more intense than the molecular transfer of heat. The thickness of the thermal boundary layer is less than the thickness of the dynamic layer, and the turbulent transfer of heat becomes marked, both in the immediate vicinity of a viscous sublayer and in the viscous sublayer. The last effect makes its distinct appearance in media with a Prandtl number $Pr > 100$.

Fluids for which $Pr \ll 1$ (all molten metals). In this class of fluids the molecular transfer of heat is considerably more intense than the molecular transfer of momentum. The thickness of the thermal boundary layer is greater than the thickness of the dynamic layer, in which connection the influence of molecular heat conduction is substantially pronounced in the turbulent core of the flow.

Connection Between the Turbulent Viscosity and Turbulent Heat Conduction In The Planar Flow of An Incompressible Fluid

It is a well-known fact that in planar turbulent flow with small pressure gradients the turbulent tangential stresses and heat flow due to turbulent heat transfer are determined with great accuracy by the expressions

$$\tau_T = -\rho\overline{V_x V_y}$$

and

$$q_T = -\rho c g \overline{V_y \theta}, \tag{4.1}$$

where ρ is the density of the fluid, c is the specific heat, g is the gravitational acceleration, V_x and V_y are the fluctuation components of the velocity vector, θ is the fluctuation component of the temperature.

From these expressions it is evident that the momentum ρV_x and the heat content $gc\rho\theta$ are referred to the same fluctuations V_y, i.e., the turbulent transfer of momentum and heat comes about through the same moving mass (mole) of the medium. From this it follows that between the coefficients of turbulent heat conduction and turbulent viscosity there should be a direct proportionality, i.e.,

$$\lambda_T = \varepsilon g c \mu_T, \tag{4.2}$$

where ϵ is a coefficient which accounts for the unlike distributions of heat content and momentum of a mole over its path of motion during the fluctuation. Transforming (4.2) to a dimensionless form, we obtain

$$\frac{\lambda_T}{\lambda} = \varepsilon Pr \frac{\mu_T}{\mu}. \tag{4.3}$$

This equation distinctly shows that as the Pr number decreases the relative effect of the turbulent heat conduction drops considerably. It is easily noted that the quantity ϵ is inversely proportional to the turbulent Prandtl criterion. In fact,

$$\frac{1}{\varepsilon} = \frac{g c \mu_T}{\lambda_T} = Pr_T. \tag{4.4}$$

In the turbulent flow nucleus of a fluid with constant physical properties we have a logarithmic velocity profile:

$$\frac{w}{v^*} = 5{,}5 + \frac{1}{0{,}4} \ln \eta, \tag{4.5}$$

where $v^* = \sqrt{\frac{\tau_W}{\rho}}$ is the so-called "tangential stress velocity" and $\eta = \frac{v^* y}{\nu}$ is a dimensionless quantity which is characteristic of the distance from the wall (y is the distance from the wall, ν is the kinematic viscosity coefficient).

In this region of flow in a circular tube,

$$\frac{\mu_T}{\mu} = \frac{\tau_T}{\mu \frac{dw}{dy}} = \frac{\tau_W\left(1 - \frac{y}{R_0}\right)}{\mu \frac{dw}{dy}} = 0{,}4\eta\xi, \tag{4.6}$$

where $\xi = \frac{R}{R_0}$ is the ratio of the flow radius to the tube radius.

The thickness of the viscous sublayer in a two-layer scheme of turbulent flow is determined by the value $\eta_1 = 11.6$. Such a scheme is insufficiently valid near the viscous sublayer, since in this it is assumed that in the region $\eta < 11.6$ the ratio $\mu_T/\mu = 0$, while in the region $\eta > 11.6$ the ratio $\mu_T/\mu \gg 1$. Here the quantity μ_T suffers a discontinuity, specifically: to the left of $\eta = 11.6$, $\mu_T = 0$, while to the right of this value of η the ratio $\mu_T/\mu = 0.4 \cdot 11.6 = 4.6$. Nevertheless, this scheme makes it possible to calculate the frictional drag with reasonable accuracy. In calculating the heat transfer in liquid metals a two-layer scheme is all the more applicable, since even for $Pr = 0.1$ the ratio μ_T/μ on the side of the turbulent core bounded by the viscous sublayer does not exceed 0.46.

The Magnitude of the Nonsimilarity Coefficient for the Transfer of Heat and Momentum

Certain properties of the quantity ϵ in equation (4.3) can be explained on the basis of the following considerations. For horizontal turbulent flow it is possible to write the familiar relation

$$-q = (\lambda + \lambda_T)\frac{\partial t}{\partial y} = \lambda \frac{\partial t}{\partial y} + c\rho g V_y \theta, \tag{4.7}$$

$$\tau = \mu \frac{\partial w_x}{\partial y} - \rho V_x V_y. \tag{4.8}$$

Here the order of the mean square of the fluctuation component of the velocity is determined by the expression $v^* = \sqrt{\frac{\tau}{\rho}}$.

From these equations it follows that the nonsimilarity coefficient for the transfer of heat and momentum in connection with the turbulent transfer factor ϵ is a function of at least six variables, viz.,

$$\varepsilon = f\{v^*,\ y,\ cg,\ \rho,\ \mu,\ \lambda\}, \tag{4.9}$$

which are composed of four fundamental dimensions (kg, sec, m, kcal/sec), i.e., they are combined in accordance with the π-theorem of dimensional analysis into two dimensionless complexes. For the latter it is convenient to assume the already familiar dimensionless quantities:

$$\eta = \frac{v^* y}{\nu}, \quad Pr = \frac{cg\mu}{\lambda}. \tag{4.10}$$

At a sufficiently great distance from the viscous sublayer ($\eta > 11.6$) in equation (4.8) it is possible to neglect the molecular viscosity and, correspondingly, to eliminate the quantity μ. Then five remaining quantities are made up of the same four dimensions and should combine into a single dimensionless complex

$$\frac{v^* y}{a} = \eta\, Pr. \tag{4.11}$$

These considerations are valid for the region of the turbulent core in which the relative distance y/δ from the walls is still not too pronounced, where δ is the thickness of the boundary layer, or the radius of the tube.

Thus, in the general case we have

$$\varepsilon = \varepsilon\left(\eta;\ Pr;\ \frac{y}{\delta}\right), \tag{4.12}$$

while in the region $y_1 \ll y \ll \delta$

$$\varepsilon \approx \varepsilon(\eta;\ Pr). \tag{4.13}$$

For all practical purposes, evidently, we can confine ourselves to the relation (4.13), which gives fair accuracy for all regions of the turbulent core.

From data on turbulent transfer in horizontal streams of a gas and of nonmetallic liquids it is well known that for $Pr > 0.5$ the coefficient $\epsilon \approx 1$. On the other hand, for $Pr = 0$ the turbulent transfer of heat is generally impossible and it is certain that in this case $\epsilon = 0$. From these considerations it follows that the function ϵ should satisfy the limiting conditions:

$$\begin{aligned} &\text{for} \quad Pr = 0 \quad \varepsilon = 0, \\ &\text{for} \quad Pr \to \infty \quad \varepsilon \to 1. \end{aligned} \tag{4.14}$$

Amplifying on the ideas of Prandtl's semi-empirical theory of turbulence [50], Voskresenskii [8, 80] obtained a simple analytic expression for the coefficient ϵ. A quite uniquely formulated derivation describing the process consists of the following.

It is assumed that: a) there exists a planar turbulent flow in which the momentum and heat are transferred by one and the same portions (moles) of the liquid which is moving as a result of turbulent fluctuations imposed on the average current; b) a mole is spherical in form; c) the propagation of heat inside and outside the mole proceeds solely as a result of molecular heat conduction due to the high thermal conductivity of the liquid metal and to the weak action of the turbulent fluctuations on the mole, which in the present case is considered as the only carrier of heat arising from turbulent perturbations in the flow; d) within the limits of motion of the mole, i.e., over the displacement path length the temperature distribution in the main flow is almost linear.

The scheme described above is shown graphically in Fig. 4.1. In the cross section 1 – 1 there arises a turbulent mole of diameter D moving across the flow with a velocity V_y. In the cross section 2 – 2, whose distance from the cross section 1 – 1 is the displacement path length l, the mole is displaced with the remaining flow.

The temperature of the mole at the instant of its origination is equal to the temperature of the main flow in the cross section 1 – 1, i.e., $\overline{t}_1 = t_{01}$. Over the path l, the process of heat transfer occurs between the mole and the remaining mass of the liquid, as a result of which the mole is carried to the cross section 2 — 2 with a temperature $\bar{t}_2 \neq \bar{t}_1$.

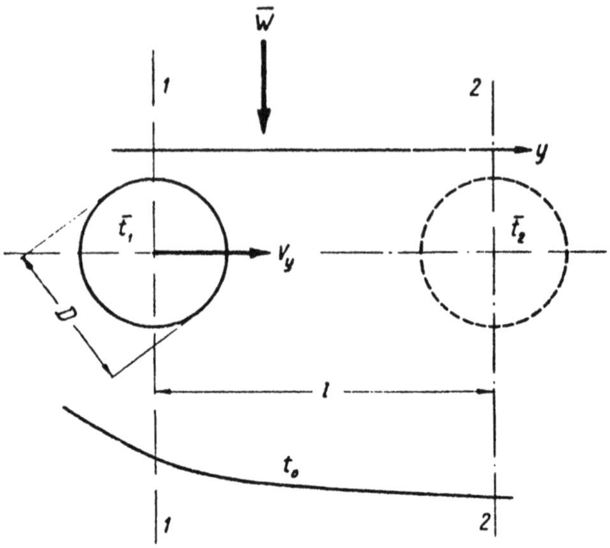

Fig. 4.1. Displacement scheme of a turbulent mole.

Since, according to Prandtl's semi-empirical theory of turbulence, in displacing over the path l the turbulent mole completely transports the surplus momentum ρV_x, assuming similarity of the transfers of momentum and heat, when $\epsilon = 1$, the mole should transport the surplus heat $c\rho g(t_{02} - t_{01})$ to the cross section 2 — 2. Consequently, the nonsimilarity coefficient for the transfer of heat and momentum can be defined as the ratio

$$\varepsilon = \frac{t_{02} - t_2}{t_{02} - t_{01}}. \tag{4.15}$$

In correspondence with assumption d) we have

$$t_{02} = t_{01} + \frac{\partial t}{\partial y} l, \tag{4.16}$$

where $\partial t / \partial y$ is the intensity of the mean temperature variation along the y-axis.

Substituting this value of t_{02} into the expression (4.15), we obtain

$$\varepsilon = 1 - \frac{\bar{t}_2 - t_{01}}{\frac{\partial t}{\partial y} l}. \tag{4.17}$$

In correspondence with assumption b) we shall consider the process of temperature variation in the displaced mole as the temperature variation in a sphere whose temperature is varying linearly with time. The solution to this problem [49], as proposed by Lykov, has the form

$$t_{02} = t_{01} + b\tau,$$

$$(\bar{t}_2 - t_{01}) \frac{4a}{bD^2} = Fo - \frac{1}{15}\left(1 + \frac{5}{Bi}\right) + \sum_{1}^{\infty} C_i \exp(-\beta_i Fo), \tag{4.18}$$

where b is the rate of change of the temperature of the medium in deg/hr, $F_0 = \frac{4a\tau}{D^2}$ is the Fourier number, $Bi = \frac{\alpha D}{2\lambda}$ is the Biot number, $a = \frac{\lambda}{c\gamma}$ is the thermal conductivity of the material of the sphere, and C_i, β_i are coefficients which are functions of the Bi number.

In correspondence with assumption c) the coefficient of heat transfer from the mole to its surrounding medium is determined by the minimum value of the Nu number for the sphere, and this, as we know, is equal to 2. On the other hand, as a consequence of the homogeneity of the substance of the mole and the remaining part of the flow, λ for the medium is equal to λ for the sphere and $2\,Bi = Nu = 2$.

The calculated time corresponding to the displacement of the mole from the cross section 1 — 1 to the cross section 2 — 2 is equal to

$$\tau = \frac{l}{v_y^*} \tag{4.19}$$

and

$$Fo = \frac{4a}{v_y^* l}\left(\frac{l}{D}\right)^2. \tag{4.20}$$

where $v_y^* = \sqrt{\frac{\tau_R}{\rho}}$ is the velocity due to tangential stress at a distance y from the wall. The rate of change of temperature over the path of displacement is equal to

$$b = \frac{t_{02} - t_{01}}{\tau} = \frac{\frac{\partial t}{\partial y} l}{\tau} \tag{4.21}$$

and

$$(\bar{t}_2 - t_{01}) \frac{4a}{bD^2} = \frac{t_2 - t_{01}}{\frac{\partial t}{\partial y} l} Fo. \tag{4.22}$$

Making use of the relations (4.18) - (4.22) and substituting the value of the relative temperature variation of the mole as determined by means of these relations into (4.17), for Bi ≈ 1, we obtain

$$\varepsilon = \frac{1}{4} Z \left[\frac{8}{15} - \sum_{i=1}^{\infty} C_i \exp\left(-\beta_i^2 \frac{4}{Z} \right) \right], \tag{4.23}$$

where

$$Z = \left(\frac{D}{l} \right)^2 \frac{v_y^* l}{a}.$$

The values of the coefficients C_i and β_i are given in Table 4.1.

TABLE 4.1
Values of the Coefficients in Equation (4.23), Assuming Bi = 1.

i	1	2	3
C_i	0,3994	0,0005	0,00003
β_i	1,5708	4,7124	7,8540

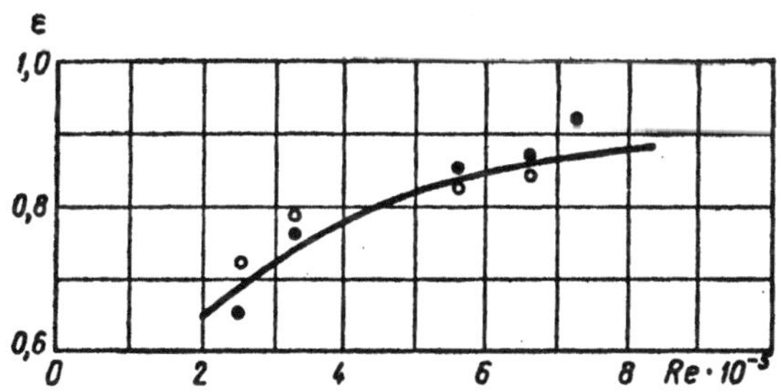

Fig. 4.2. Variation in the nonsimilarity coefficient for the turbulent transfer of heat and momentum over an average radius ($y/R_0 = 0.5$) as a function of the Re number. The different kinds of points are used to designate the ascending and descending directions of flow of mercury.

The values of the coefficients C_i in equation (4.3) diminish very rapidly, since we can limit ourselves to the first term only of the series. Furthermore, rounding off the values of C_i to 0.4 and that of β_i^2 to 2.5, we obtain a simple analytic expression for the coefficient ε:

$$\varepsilon = \frac{Z}{10} \left(1 - e^{-\frac{10}{Z}} \right). \tag{4.24}$$

From hydrodynamics it is a familiar fact that

$$v_y^* = v^* \sqrt{1 - \frac{y}{R_0}} = \bar{w} \sqrt{\frac{\zeta}{8} \left(1 - \frac{y}{R_0} \right)}, \tag{4.25}$$

where $v^* = \sqrt{\frac{\tau_w}{\rho}}$ is the velocity due to horizontal stress on the wall of the tube, ζ is the frictional drag coefficient, y is the distance from the wall of the tube, and R_0 is the radius of the tube.

Moreover, in the region $y_1 < y \ll R_0$, we have

$$l \approx 0,4y. \tag{4.26}$$

Making use of the relations (4.25) and (4.26), we obtain

$$Z = 0{,}4\left(\frac{D}{l}\right)^2 Pe\sqrt{\frac{\zeta}{32}\left(1-\frac{y}{R_0}\right)\frac{y}{R_0}}, \qquad (4.27)$$

where $Pe = \frac{\bar{w}2R_0}{a}$ is the Peclet number.

The ratio D/l in a growing turbulent flow weakly depends on the Re number and is equal to approximately 1.

Introducing the notation

$$Z_1 = Pe\sqrt{\frac{\zeta}{32}\left(1-\frac{y}{R_0}\right)\frac{y}{R_0}}, \qquad (4.28)$$

$$B = 0{,}04\left(\frac{D}{l}\right)^2.$$

we rewrite equation (4.24):

$$\varepsilon = BZ_1\left(1 - e^{-1/BZ_1}\right). \qquad (4.29)$$

In doing this, strictly speaking, the coefficient B should be considered as an experimentally determined constant.

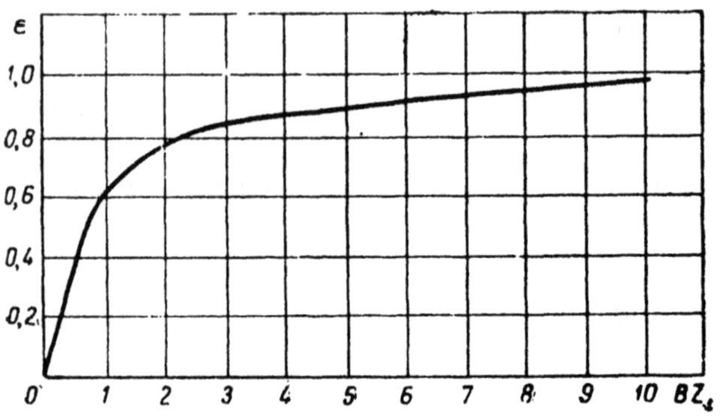

Fig. 4.3. Behavior of ϵ as a function of Z_1 for B = 0.032.

In Fig. 4.2, the analytic value of ϵ, calculated according to this equation for B = 0.032 are compared with the values of ϵ determined experimentally in the flow of mercury through a circular tube [12]. The fully satisfactory agreement of the experimental data with the theoretical is clearly visible.

In Fig. 4.3, the variation of the coefficient ϵ as a function of the magnitude of the complex Z_1 is shown.

It is evident that for large value of Z_1 the coefficient ϵ is practically equal to unity, which also takes place in gases and nonmetallic liquids.

Chapter 5

HEAT TRANSFER DURING FLOW IN TUBES

Theoretical Solutions

The fundamental equation of heat conduction during the flow of an incompressible liquid in a circular tube is usually written as [6]:

$$\frac{\partial}{\partial R}\left[R(\lambda+\lambda_{\text{T}})\frac{\partial t}{\partial R}\right]=c\gamma w R\frac{\partial t}{\partial x}, \tag{5.1}$$

where t and w are the point values of the temperature and velocity of flow, averaged during a period sufficiently large in comparison with the period of turbulent fluctuation.

It is possible to consider two basic boundary conditions: the constancy of the heat flow density at the surface of the tube (q = const) and the constancy of the temperature of the surface of the tube (t_W = const).

When q = const the longitudinal component of the temperature gradient, when averaged over the flow cross section, also remains constant and is equal to

$$\frac{\overline{\partial t}}{\partial x}=\frac{2q}{c\gamma\overline{w}_0 R_0}, \tag{5.2}$$

where $\overline{w}_0$ is the average mass flow rate of the liquid, R_0 is the inside radius of the tube ($2R_0 = D$).

If in equation (5.1) the quantity $\partial t/\partial x$ is replaced by the quantity $\overline{\partial t}/\partial x$ in accordance with equation (5.2), then to a first approximation we can obtain an integral equation which is suitable for the calculations [11]:

$$Nu=\left[2\int_0^1\frac{\left(\int_0^{\xi}\omega\xi\,d\xi\right)^2}{\left(1+\varepsilon Pr\frac{\mu_{\text{T}}}{\mu}\right)\xi}d\xi\right]^{-1}, \tag{5.3}$$

where $Nu = \frac{\alpha D}{\lambda}$ is the Nusselt number, $\xi = \frac{R}{R_0}$ is the ratio of the flow radius to the radius of the tube, and $\omega = \frac{w}{\overline{w}}$ is the ratio of the velocity at the given point to the average mass flow rate.

The equation gives satisfactory results in the region of stabilized heat transfer, i.e., at a sufficient distance from the tube input.

In the case of laminar flow $\mu_T = 0$ and

$$\omega = 2(1-\xi^2). \tag{5.4}$$

Substituting these values of ω and μ_T into the expression (5.3), we find that for the region of stabilized heat transfer in a circular tube, assuming laminar flow, Nu = 4.36.

A point solution of (5.3) for $t_w = \text{const}$ gives a value $Nu = 3.66$ for the region of thermal stabilization, i.e., it is sufficiently near the value obtained above.

The velocity profile and, correspondingly, the turbulent viscosity coefficient exhibit almost no dependence on the thermal characteristics (the effect of heat flow on the heat transfer and frictional drag are considered below). Consequently, the Nu number is a function only of thermal characteristics such as the Pr number and the coefficient ϵ. In accordance with (5.3), the decrease in Pr and the decrease in ϵ connected with this lead to a decrease in Nu, i.e., to a reduction in the intensity of heat transfer. Thus, in the turbulent flow of a liquid and for a given number $Re = \frac{\overline{w}D}{\nu}$ the minimum value of the dimensionless heat-transfer coefficient occurs when $Pr = 0$.

In the interval of the Re number from $5 \cdot 10^3$ to $2 \cdot 10^5$ the logarithmic velocity profile [see equation (4.5)] is in good approximation to the equation

$$\omega = 1{,}22\xi^{1/7}. \tag{5.5}$$

Substituting this value of ω into (5.3) when $Pr = 0$, we obtain $Nu = 6.8$.

Consequently, in stabilized turbulent flow when $Pr \to 0$ the quantity Nu tends to some constant value, which is greater than in the case of stabilized laminar flow.

If we neglect the variation in velocity in the immediate vicinity of the solid wall, then in the remaining portion of the turbulent flow the velocity is little different from its average steady flow value. Substituting the quantity $\omega = 1$ (linearized velocity profile $w = \overline{w}$) into (5.3), when $Pr = 0$ we obtain $Nu = 8$, which is sufficiently near the point value $Nu = 6.8$.

A numerical solution of (5.1) for $q = \text{const}$ was obtained by Martinelli [10] and Lyon [11, 21]. The latter showed that the analytic results obtained are, to a definite degree of proximity, in good approximation to the following equation in the interval of the Pr number characteristic for liquid metals:

$$Nu \approx 7 + 0{,}025\, Pe^{0,8}, \tag{5.6}$$

where $Pe = \frac{\overline{w}D}{a}$ is the Peclet number.

Seban and Shimazaki [55] performed calculations for $t_w = \text{const}$, $\epsilon = 1$ and obtained the equation

$$Nu \approx 5 + 0{,}025\, Pe^{0,8}. \tag{5.7}$$

Thus, the condition $t_w = \text{const}$ leads to a somewhat lesser value of the constant terms than is the case when $q = \text{const}$.

Values of the number Nu for the flow of liquid metals in a tube when the variation in the coefficient ϵ is taken into account were calculated by Voskresenskii from equation (4.24). The results of these calculations are approximated by the equation

$$Nu \approx 5 + 0{,}014\, Pe^{0,8}. \tag{5.8}$$

Equation (5.3) can be solved in finite form, if we let $\omega = 1$ and $\epsilon = \overline{\epsilon} = \text{const}$, where $\overline{\epsilon}$ is some mean effective value of the nonsimilarity coefficient for the turbulent transfer of heat and momentum. In this case for a two-layer scheme of turbulent flow (5.3) can be written as follows:

$$Nu \approx 2\left(\int_0^{\xi_1} \frac{\xi^3\, d\xi}{1 + P(1-\xi)\xi} + \int_{\xi_1}^{1} \xi^3\, d\xi\right)^{-1}, \tag{5.9}$$

where $P = 0{,}4\epsilon Pe \sqrt{\frac{\zeta}{32}}$.

Here it is assumed that in the flow core, in correspondence with the logarithmic profile of the velocity distribution, the turbulent viscosity is determined from an equation given in [6]:

$$\frac{\mu_T}{\mu} = 0{,}4\,Re\sqrt{\frac{\zeta}{32}}\,(1-\xi)\,\xi, \tag{5.10}$$

where ζ is the frictional drag coefficient.

The calculated boundary of a viscous sublayer is determined from the equation

$$\xi_1 = 1 - \frac{11{,}6}{Re}\sqrt{\frac{32}{\zeta}}\,. \tag{5.11}$$

Integrating (5.9) with ϵ = const and applying the relations (5.10) and (5.11), we obtain

$$Nu \approx 2P\left[\frac{P+3}{2\sqrt{P^2+4P}}\ln\frac{(\sqrt{P^2+4P}+P+0{,}8\bar{\epsilon}Pr)(\sqrt{P^2+4P}+P)}{(\sqrt{P^2+4P}-P+0{,}8\bar{\epsilon}Pr)(\sqrt{P^2+4P}-P)} - \frac{P+1}{2P}\ln(1+4{,}63\bar{\epsilon}Pr) - \frac{3}{2} + 4{,}63\bar{\epsilon}Pr\right]^{-1}. \tag{5.12}$$

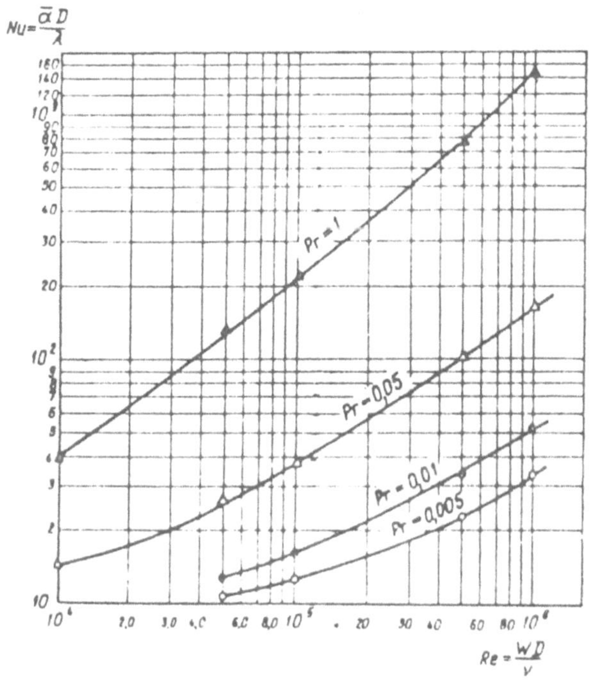

Fig. 5.1. The relation Nu = f (Re) for $\bar{\epsilon}$ = 1.

In Fig. 5.1 the Nu number is shown as a function of the Re and Pe numbers, as calculated according to (5.12) when ϵ = 1. The behavior of these curves remains invariant even for other values of $\bar{\epsilon}$. It is evident that the usual dependence of Nu = f (Re) is the case of nonmetallic media begins in the case of liquid metals to experience the extremely powerful influence of a second determining criterion, namely the Pr number.

In Fig. 5.2 the results of calculations according to (5.12) for various values of $\bar{\epsilon}$ and Pr, characteristic for liquid metal heat-transfer media, are shown.

The quantity $\Delta Nu = Nu - 8$, which in the present case shows an increase in the intensity of heat transfer as a result of the turbulent transfer of heat, is plotted on the vertical axis. The complex P is plotted on the horizontal.

It is evident that the complex equation (5.12) is well approximated by the simple relation

$$Nu \approx 8 + 1{,}3P^{0{,}81}. \tag{5.13}$$

Making somewhat of a correction to this result, in correspondence with the value Nu = 6.8, we obtain [54]

$$Nu \sim 6{,}8 + 1{,}3P^{0{,}81}. \tag{5.14}$$

Substituting here the expression

$$\zeta = 0{,}184Re^{-0{,}2}, \tag{5.15}$$

which is valid for smooth tubes in a broad interval of Re, we obtain

$$Nu \approx 6{,}8 + 0{,}0765\bar{\epsilon}^{0{,}81}Pr^{0{,}081}Pe^{0{,}73}. \tag{5.16}$$

From this it follows that the Prandtl criterion for numbers Pr ≪ 1 very slightly influences the relationship between the numbers Nu and Pe.

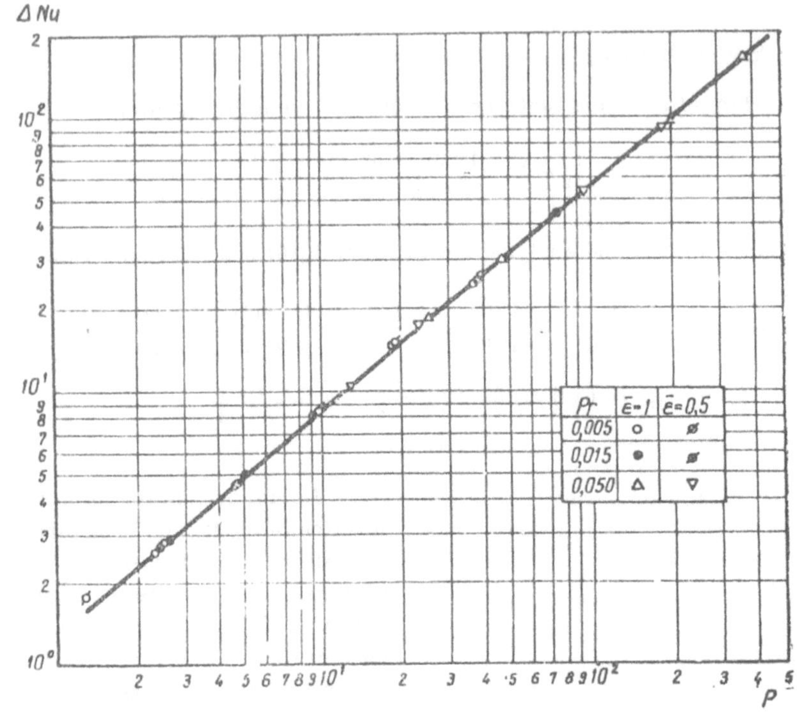

Fig. 5.2. The relation $\Delta Nu = f(P, Pr)$, where $P = 0{,}4\bar{\varepsilon} Pe \sqrt{\frac{\zeta}{32}}$.

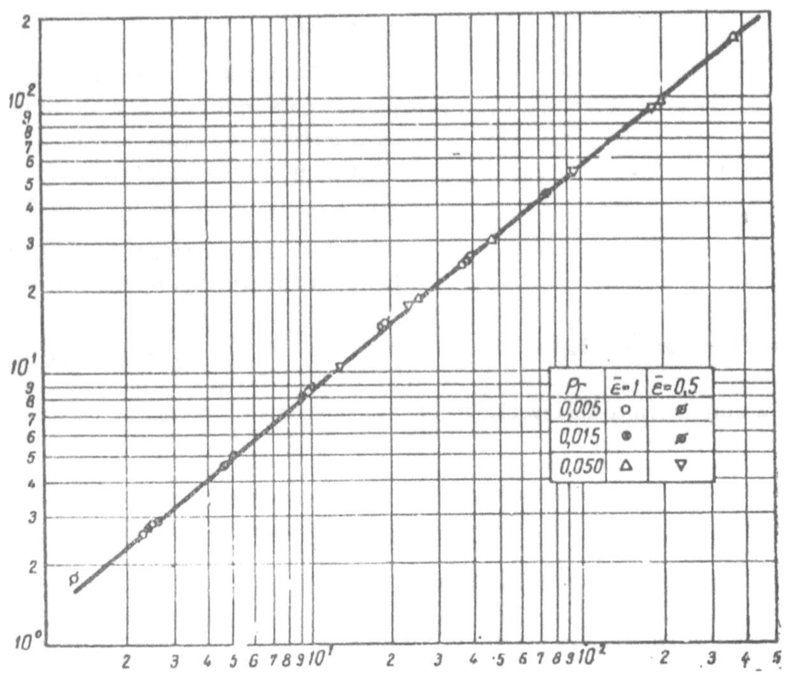

Fig. 5.3. Comparison of calculations according to (5.16) with experimental data on heat transfer in eutectic lead-bismuth (Pr ≈ 0.025).

This means that if in the general case for heat transfer in tubes the criterion function

$$Nu = f\left(Pe; \ Re; \ \frac{L}{D} \right),$$

holds, or, what amounts to the same thing,

$$Nu = f\left(Re; \ Pr; \ \frac{L}{D} \right), \tag{5.17}$$

then for liquid metals in both the laminar and turbulent case

$$Nu \approx f\left(Pe; \ \frac{L}{D} \right). \tag{5.18}$$

The form of this function depends on the conditions of flow.

Figure 5.3 gives an illustration of the results of experiments performed by Borishanskii, Kutateladze, and Shneiderman [5], [33] and Johnson, Clabaugh, and Hartnett [18]. Curve 1 is calculated from (5.16) for $\bar{\epsilon} = 0.33$. It is evident that the experimental points lie near the theoretical curve corresponding to Nu as a function of Pe; thus, we must consider the fact that in liquid metals the turbulent transfer of heat is considerably weaker than the transfer of momentum ($\bar{\epsilon} < 1$). Curve 2 in the figure is calculated according to (5.16), again for $\bar{\epsilon} = 0.33$, but with the constant term equal to 5.0 (for $t_W = \text{const}$). This curve corresponds better with the data in the region of fairly small Pe numbers.

Since, in addition to conditional factors, the intensity of heat-transfer with liquid metals depends further on a whole series of other stipulations (the purity of the metal, its interaction with the wall, etc.), the effect of which is still not too clearly established, some experimental data on certain basic liquid metal heat-transfer media are given below.

Experimental Data on Heat-Transfer To Mercury

The earliest papers on heat-transfer in the flow of liquid metals in tubes were those of Styrikovich, Sorin, and Semenovker [1, 2]. In these heat-transfer to mercury (t = 450-500° C) flowing through ordinary seamless, as well as low-alloy pipes having a diameter from 16 to 50 mm was investigated; in the investigated interval of the numbers $Re = (150\text{-}350) \cdot 10^3$ these pipes should be considered rough. The apparatus consisted of a naturally-circulating vertical coil. The pipes were heated in radiation ovens. The results of the experiments for pipes having four different diameters are shown in Fig. 5.4. As evident from the graph, the experimental points are distributed about a line which gives a variation in the Nu number from 7 to 11 for a variation in Pe number from 600 to 1800.

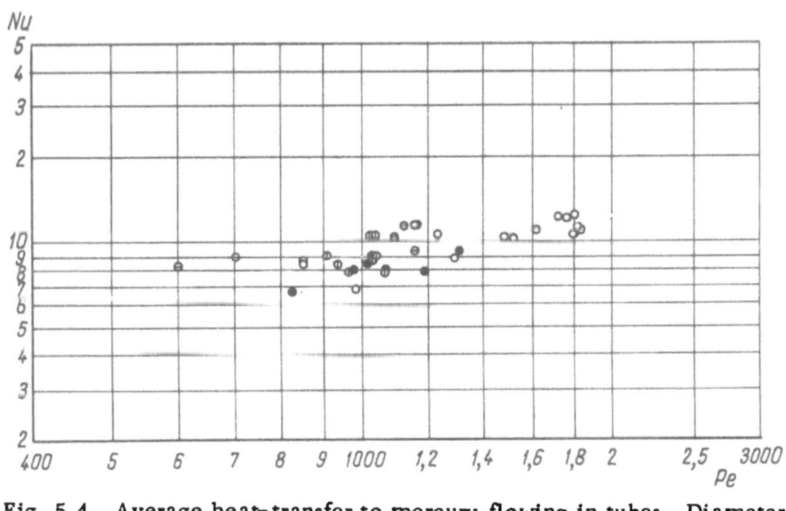

Fig. 5.4. Average heat-transfer to mercury flowing in tubes. Diameter in mm: ⦶) 16; ○) 32; ●) 40; ⊖) 42.

In these same papers heat-transfer to mercury flowing in an annulus with only the outside perimeter being heated ($D_H = 48$ mm, $D_B = 28$ mm) was investigated. The test results, reduced according to the hydraulic equivalent diameter, are shown in Fig. 5.5.

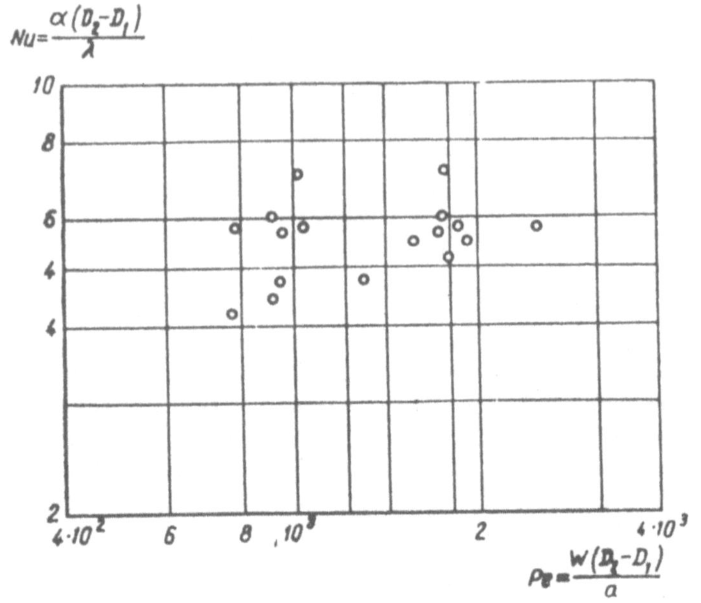

Fig. 5.5 Average heat-transfer to mercury flowing in an annulus.

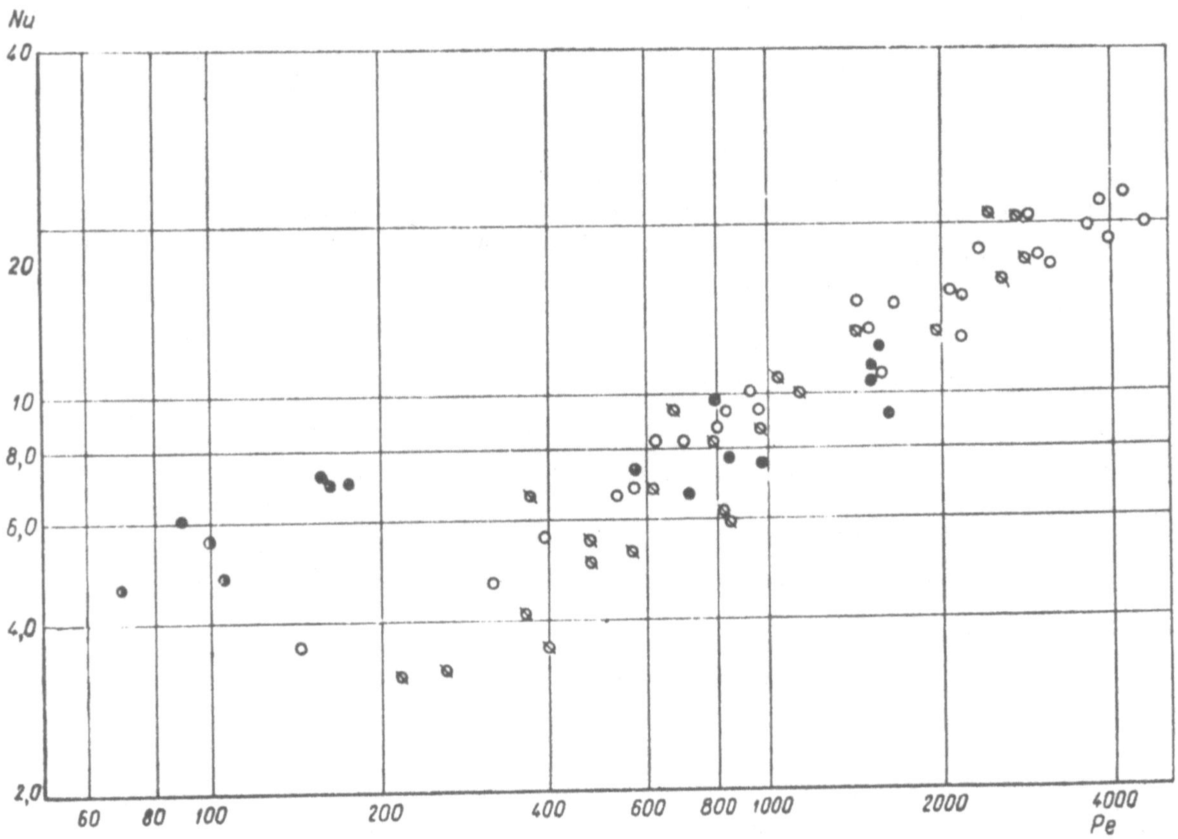

Fig. 5.6. Average heat-transfer to mercury and a mercury amalgam flowing in tubes. ○⦸) mercury; ◑●) magnesium amalgam of mercury.

Figure 5.6 shows data from the experiments of Korneev [14] with a magnesium-mercury amalgam and mercury. The experiments with the amalgam were conducted in a naturally circulating vertical coil (pipes having diameters of 28 and 40 mm were made from ElalT steel). The temperature interval was 350-500° C. The values of the Pe number varied over the interval 43-1000, while the Re number varied over the interval 8000-190,000. The experiments with mercury were conducted in pipes having diameters of 19 and 28 mm, in a forced-circulation coil at flow temperatures of 60-180° C, Pe ≈ 140-4500 and Re ≈ 8000-300,000. These data are in complete agreement both in their own self-consistency and with the data of Styrikovich, Sorin, and Semenovker.

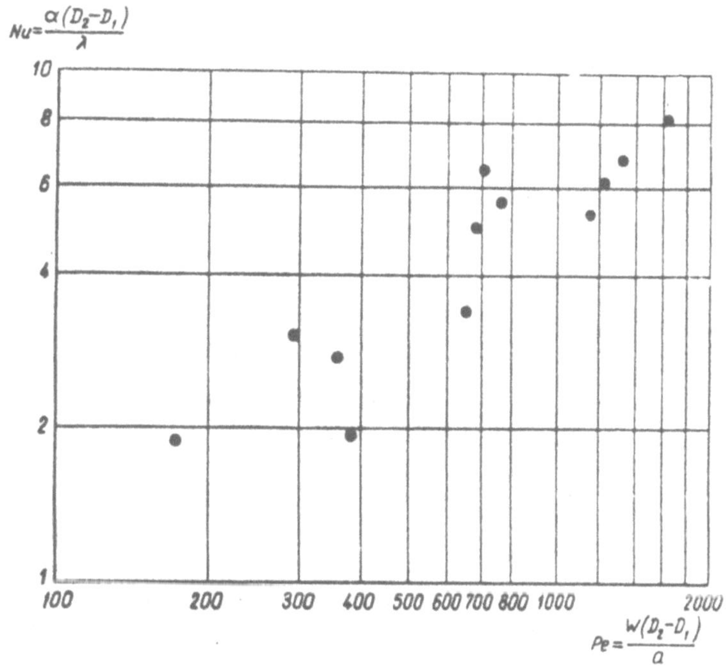

Fig. 5.7. Average heat-transfer to mercury flowing in an annulus. $D_1 = 14$ mm, $D_2 = 19$ mm.

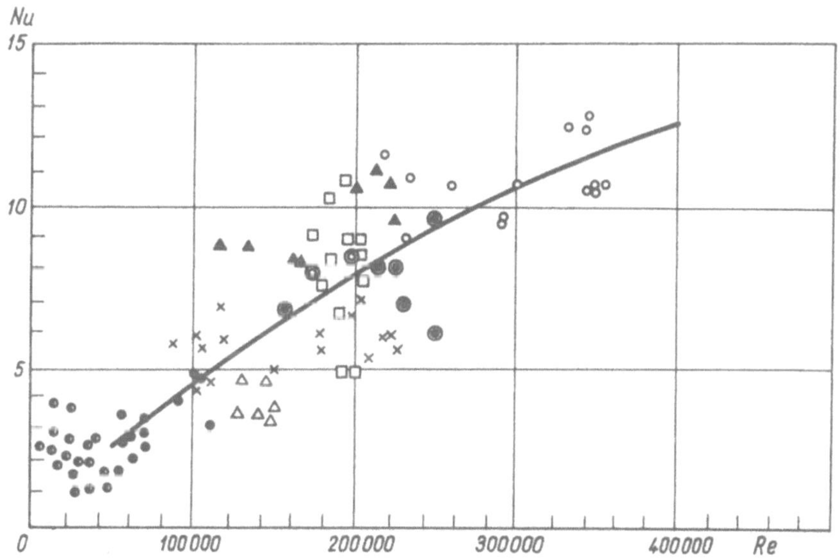

Fig. 5.8. Composite of past experiments with mercury. The various points distinguish the experiments with different tubes.

The test results with an annulus are shown in Fig. 5.7.

In Fig. 5.8 various experimental data, as reduced by Kanaev [4], for mercury are shown, along with a curve calculated from the following equation (which is valid only for values of Pr ≈ 0.005 – 0.006):

$$Nu = 0{,}0009\, Re^{0{,}74}. \tag{5.19}$$

Figure 5.9 shows the results of experiments by Voskresenskii, Kondrat'ev, Kalakutskaya, Petrov, and Osipov [7, 76] on the average coefficient of heat-transfer, obtained in the Mikheev laboratory. In these tests the diameters of the tubes (made from carbon steel) varied from 4.31 to 10.62 mm, the relative lengths (L/D) varied from 50 to 80; the mean temperature of the mercury in the tests varied over the interval 20–200° C. The ranges for the appropriate dimensionless numbers were specified as follows: Pe from 250 to 13,000; Re = 10,000 to 630,000. The tests were conducted with heated mercury, assuming a constant heat supply over the length of the tube (q = const).

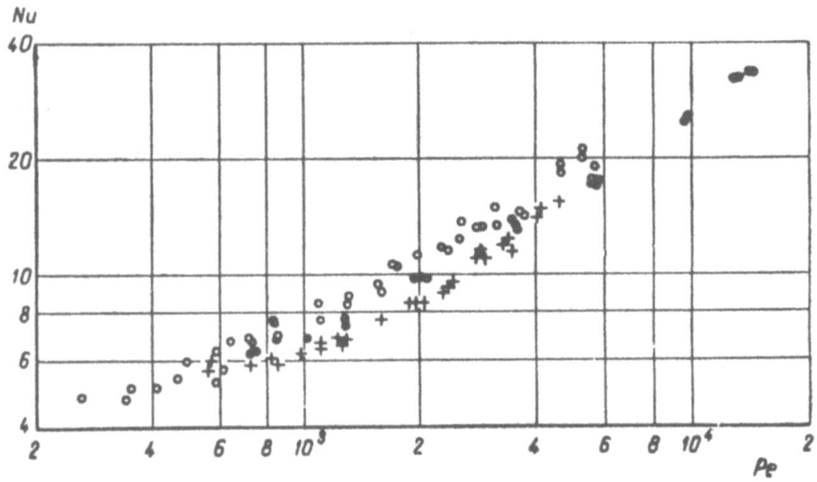

Fig. 5.9. Average heat-transfer to mercury flowing in tubes. ○) D = 4.3 mm, L/D = 80, q = const; ●) D = 10.6 mm, L/D = 64, q = const; +) D = 10.3 mm, L/D = 51 ("tube-in-tube" heat-exchanger, heating by hot water).

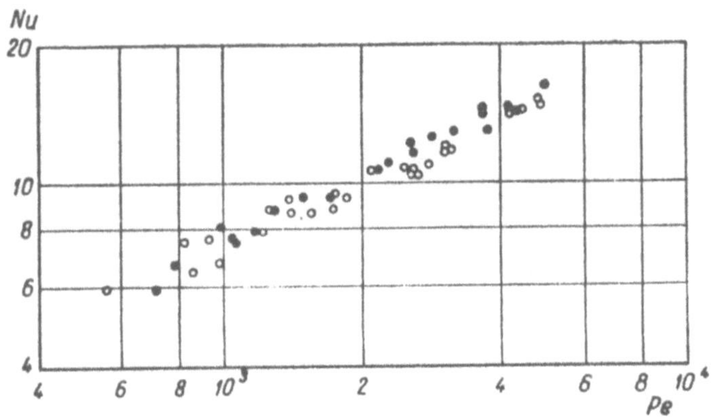

Fig. 5.10. Average heat transfer to mercury when the wall temperature of the tube is constant. ○) D = 10 mm, L/D = = 30; ●) D = 8.2 mm, L/D = 37.

The results of Fedynskii's experiments with a constant temperature of the wall of a vertical tube (t_W = = const) are shown in Fig. 5. 10. The diameters of the tubes were equal to 8.2 and 10 mm, while L/D was equal to 37 and 30, respectively. The tests were performed at a mean flow temperature of approximately 80° C and the Re number was varied from 25,000 to 100,000.

Under the same conditions an investigation of the mean coefficient of heat-transfer was made for mercury flowing in an annulus with a width δ = 2.2 mm and length L = 300 mm; D_2/D_1 = 1.54 (see Fig. 5.11). The investigation was made by a heat-exchanger method. A calculation of similarity criteria was made on the basis of a hydraulic equivalent diameter $D_e = 2\delta$.

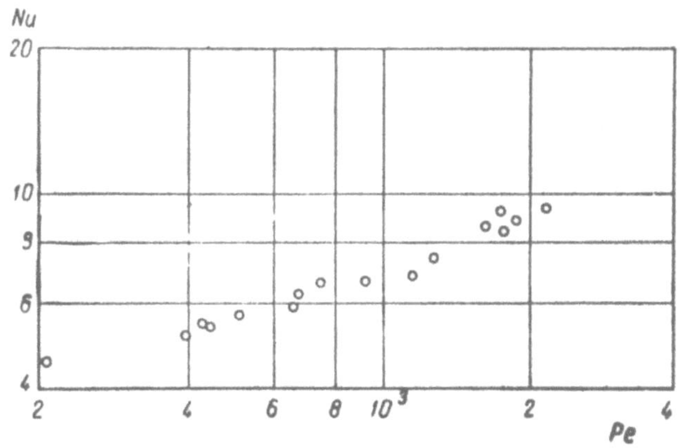

Fig. 5.11. Average heat-transfer to mercury in an annulus.

The heat-transfer in a flow of mercury in short tubes made of stainless steel for an almost constant wall temperature was investigated by Kondrat'ev [75]. The experiments were performed with tubes having a diameter of 5.3 mm and with L/D equal to 4.6; 6; 9.4; 18.9; 28.6; 38.4; and 47.7. Figure 5.12 shows the variation in the relative intensity of heat transfer during the flow of mercury in short tubes (L/D < 40). As evident from the figure, for L/D = 30 the effect of the initial portion is hardly pronounced at all and almost vanishes when L/D > > 35.

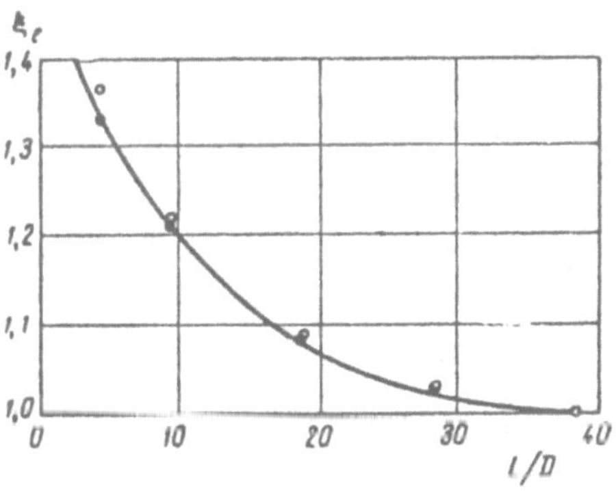

Fig. 5.12. Relative variation in heat transfer during the flow of mercury in short tubes. O) Pe = 200, ●) Pe = = 2000.

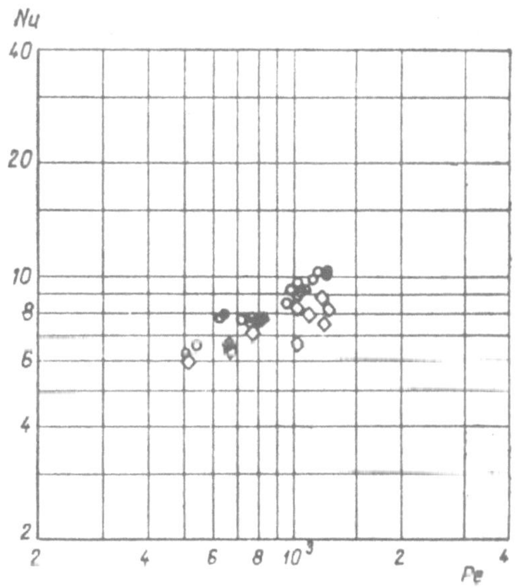

Fig. 5.13. Average heat transfer to mercury flowing in a tube. O) Heating; ◇) cooling.

The variation in heat transfer in a mercury-water heat-exchanger was investigated by Gilliland, Musser, and Page [19]. The mercury was allowed to flow in a nickel tube. The experiments were made with heated and cooled mercury. The inside diameter of the tube was 2.5 mm, L/D for the case of cooling was equal to 45, while in the case of heating it was equal to 160. Direct measurements of the wall temperature were not made and the heat-transfer coefficient had to be calculated for each medium from the heat transmission between the two, taking into account the thermal resistance at the wall and between the outer wall and the vapor or water. In these tests there was a considerable jump, amounting to nearly 100° C over-all, in the temperature along the tube. The mean temperature of the mercury flow was about 100° C. The Pe number varied within the interval 470 - 1200, while the number Re ≈ 29,000 - 70,000. The results of the tests are shown in Fig. 5.13. Somewhat of a stratification was observed at the heating and cooling points.

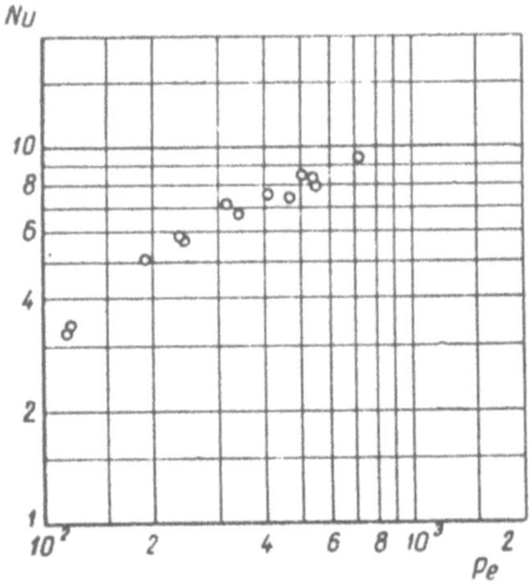

Fig. 5.14. Heat transfer beyond the stabilization length during the flow of mercury in a tube.

The local and average heat transfer to mercury, flowing in steel and nickel tubes having a diameter of 1.3 mm and L /D = = 38 were investigated by English and Barrett[15]. The tubes were heated by an electric current which flowed through an outer copper plating. This corresponded to the condition q = const along the length of the tube.

In these tests a Reynolds number interval Re = 4000 = 45,000 was investigated, and the wall temperature of the mercury did not exceed 100° C. The results of the tests on local heat transfer beyond the stabilization length are shown in Fig. 5.14. The experimental points do not lie on a horizontal, but for the numbers Pe < 300 a sudden drop in the heat transfer is observed. In Fig. 5.15 the results of this work for local heat exchange at the inflowing portion as a function of the criterion Pe•D/x, are shown, where x is the distance from the tube entrance to some cross section under investigation.

The heat transfer in mercury which is flowing over a working section, constructed in the form of a counterflowing heat-exchanger, was investigated by Trefethen [20]. The flow was investigated in a tube and in an annulus. A measurement of the wall temperature inside the tube was not made. The temperature of the outside surface of the jacket was measured. The coefficient of heat transfer was calculated on the basis of the measured heat transmission. The Re number in the tests amounted to 4600 - 102,000 for tubes having a circular cross section, the mean flow temperatures were 20-60° C, which corresponded to values of the number Pr = 0.02 - 0.0246. The experiments were conducted with copper and stainless steel (three) tubes with inside diameters of 8-20 mm and L/D = 50 to 130. The gap width δ of the annulus varied from 1 to 6 mm. A calculation of the heat transfer was made on the basis of the measured mixed mean temperature.

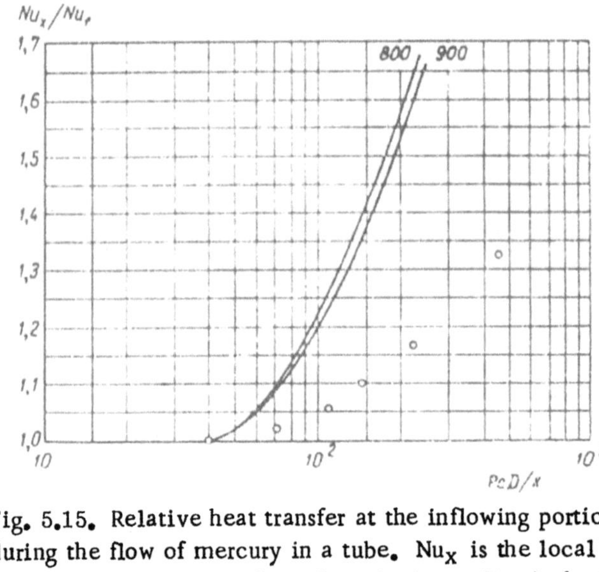

Fig. 5.15. Relative heat transfer at the inflowing portion during the flow of mercury in a tube. Nu_x is the local heat transfer at a distance x from the tube input, Nu_f is the local heat transfer beyond the stabilization length; O) experiments in which Pe = 800-900, —) analytic curves for Pe = 800 and 900.

The results of the tests on heat transfer to mercury flowing in a tube beyond the stabilization length are shown in Fig. 5.16, those for the average heat transfer are shown in Fig. 5.17, while those for an annular conduit are shown in Fig. 5.18.

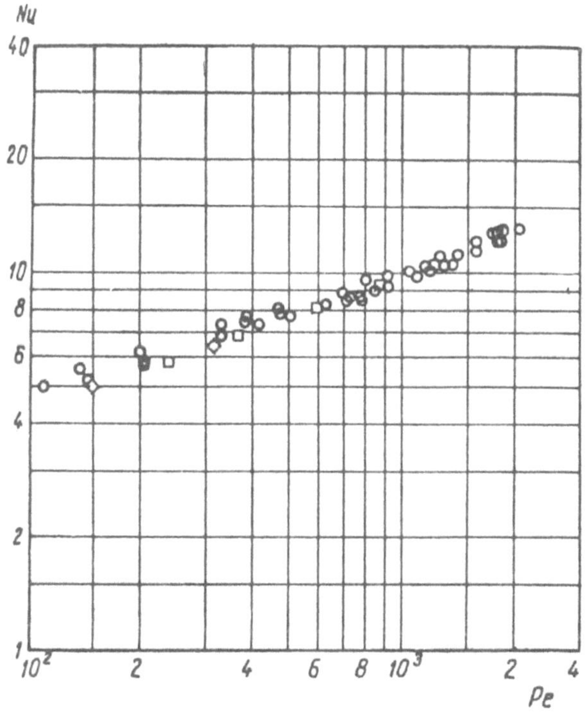

Fig. 5.16. Heat transfer to mercury flowing in a tube beyond the stabilization length.

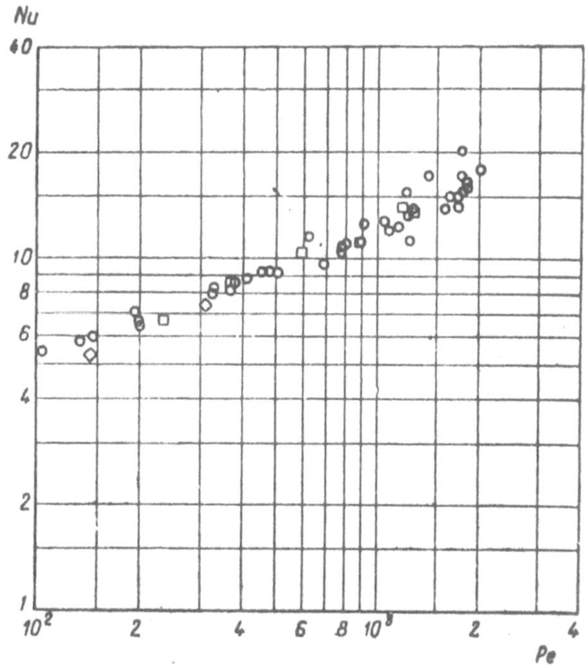

Fig. 5.17. Average heat transfer to mercury flowing in a tube.

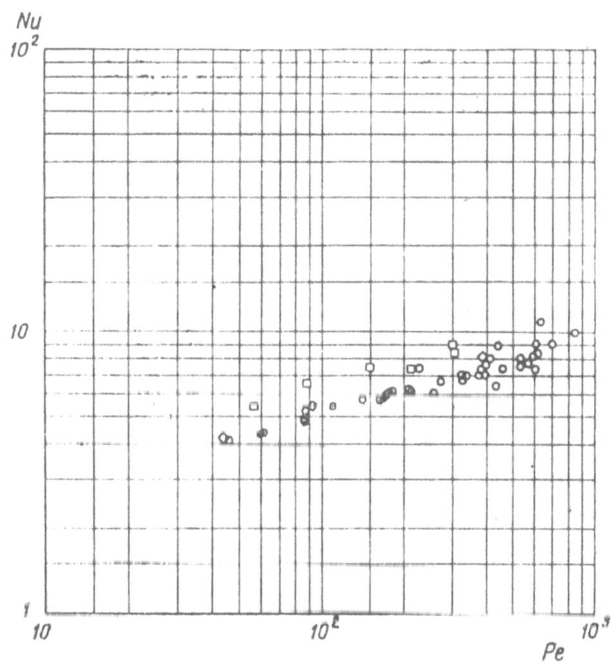

Fig. 5.18. Average heat transfer to mercury flowing in an annulus.

The heat transfer to mercury and a mercury amalgam with sodium were studied by Doody and Younger [16]. The tests were made with heating and cooling of the mercury, which was flowing in a steel tube of diameter 12.5 mm and length 1550 mm (L/D = 125). Pure mercury was used, with the numbers Re = 4000 - 24,000, temperatures 18-65° C and Prandtl numbers Pr = 0.0228 - 0.0258.

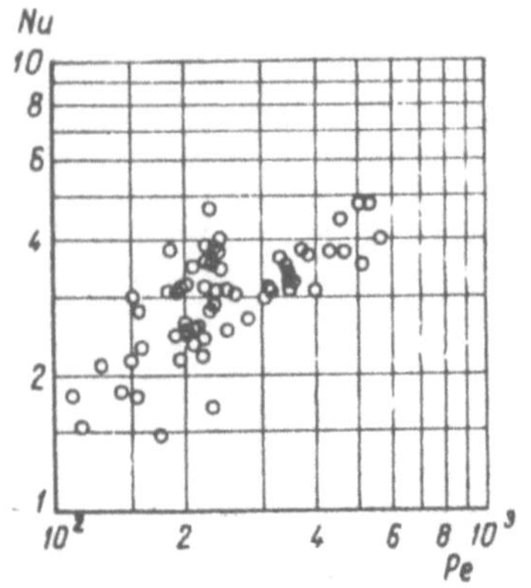

Fig. 5.19. Average heat exchange to pure mercury flowing in a tube.

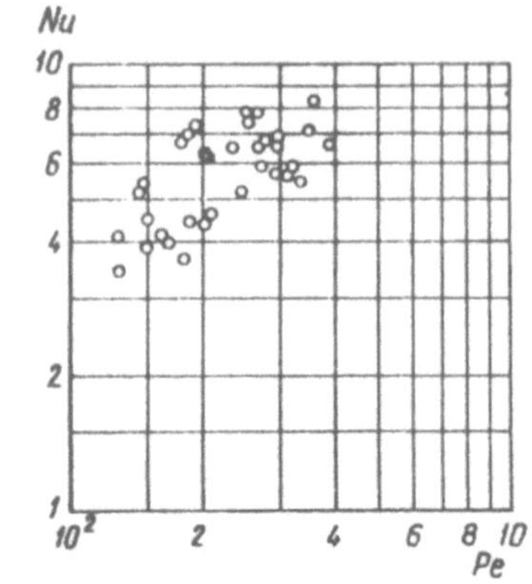

Fig. 5.20. Average heat transfer during the flow in a tube of mercury with the addition of sodium.

The tests with additions of sodium were performed with sodium concentrations of 0.008 - 0.096% (by weight). The flow temperature was 15-44° C, the Reynolds number was Re = 5600-16,000. The results of these experiments are shown in Fig. 5.19 and 5.20. A comparison of the data shows that upon the addition of sodium the heat transfer is reduced by a factor of two.

Bailey, Cope, and Watson [21] measured the heat transfer with mercury flowing in a tube made from carbon steel and having a diameter of 11 mm. The working section was made up of a heat-exchanger, in which the wall temperature was measured, making it possible to calculate the heat transfer directly. The temperature of the mercury was measured at the entrance and exit in mixing chambers and at the axis of the tube. The heat transfer was measured beyond the length of thermal and hydrodynamical stabilization at a distance from the tube entrance. The results of the experiments are shown in Fig. 5.21. In the region where Pe < 400 the heat transfer was observed to decline.

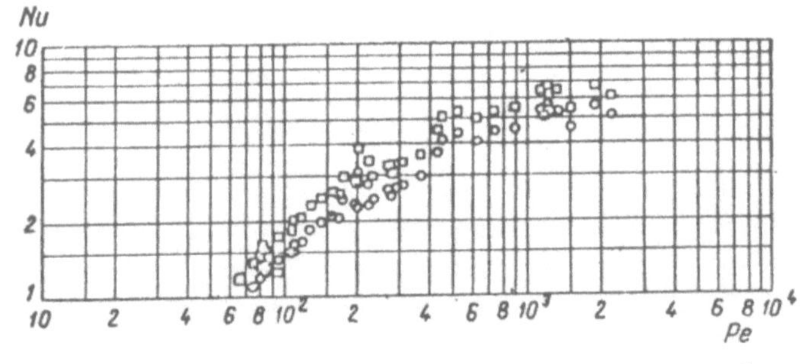

Fig. 5.21. Heat transfer beyond the stabilization length during the flow of mercury in a tube. O) Calculated according to the mixed mean temperature; □) calculated according to the tube centerline temperature of the mercury.

The heat transfer to mercury upon cooling (Fig. 5.22) was measured by Elser [21] in tubes having diameters of 6.6, 7.82, and 8.05 mm (the first two were stainless steel, the third was carbon steel). The lengths of the sections were equal to 960 mm. The wall temperature was measured at two points situated beyond the stabilization length at a distance greater than 260 mm (L/D ≈ 32 - 43.3). The temperature drop was not calculated from the mixed mean temperature, but from the centerline temperature.

Sineath [21] investigated the heat transfer to mercury in a rectangular canal. The working section was made up of two flat canals with a common wall. Heat was delivered to the mercury in one of the canals and carried off in the other. The width of the canal was 6.35 mm, its length was 635 mm (L/D = 50). The liquid temperature used in the calculations was roughly the mixed mean temperature. The measurements were made at the entrance and exit of the canals.

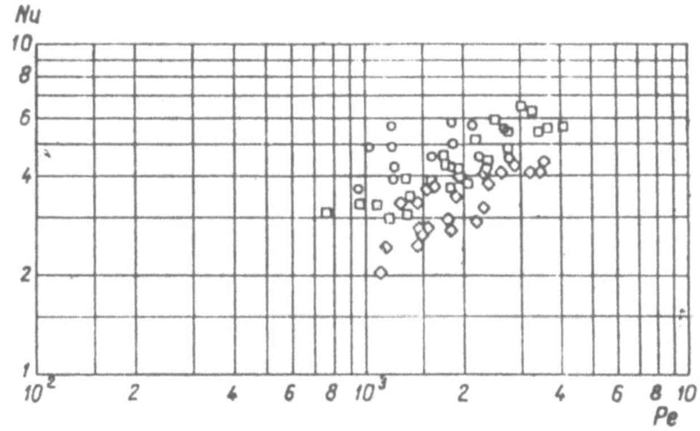

Fig. 5.22. Heat transfer beyond the stabilization length during the flow of mercury in a tube. ○) D = 8.05 mm, ◇) D = 7.81 mm, □) D = 66 mm.

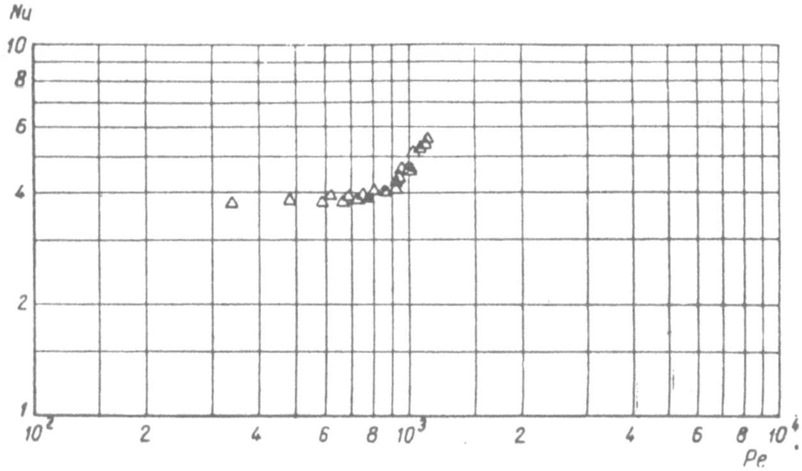

Fig. 5.23. Average heat transfer to mercury flowing in a rectangular canal.

The data obtained (Fig. 5.23) falls quite near the theoretical results for the single-wall heating of a conduit.

Isakoff and Drew [17] investigated the heat transfer in a flow of mercury in a tube (diameter 38 mm, length 5.65 m, wall thickness 3.22 mm) made from stainless steel. The outside wall temperature was measured. The temperature and velocity fields were measured at three cross sections at relative distances x/D approximately equal to 58, 98, and 138. Heat-transfer stabilization set in at the two last positions. The heat transfer was higher for x/D ≈ 58, i.e., evidently the value of the heat-transfer coefficient had not become stabilized. The temperature of the inner surface of the tube was determined by two methods: by extrapolation of the temperature profile of the liquid at the wall and by calculation according to the external temperature with allowance for losses at the wall. In the majority of the tests calculation by the second method gives a higher value for the temperature of the inside surface of the tube. The authors explain this fact by the presence of a certain thermal resistance at the interface between the liquid and solid.

The results of experiments for local heat transfer with x/D ≈ 138 are shown in Fig. 5.24, and the results of calculations of the average heat transfer during the flow of mercury in a tube are shown in Fig. 5.25.

A measurement of the heat transfer to pure mercury having additions of sodium was made by Stromquist [21], using steel tubes of varying lengths and diameters and heating the tube directly by an electric current. The temperature of the outer surface of the tube was measured. Tubes of the following dimensions were used:

Inside diameter (mm)	9.65	12.4	19.1	20
Outside diameter (mm)	19.1	25.4	38.1	38.2
Length (m)	1.2	1.22	1.27	1.27
L/D ratio	120	98.5	66.5	63.5

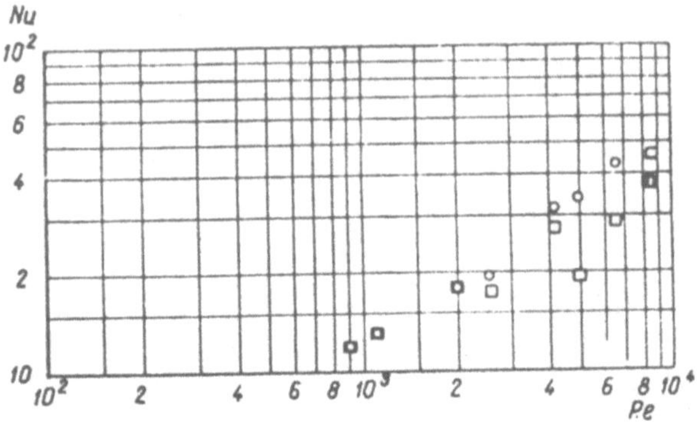

Fig. 5.24. Heat transfer beyond the stabilization length during the flow of mercury in a tube. The temperature of the inner surface was calculated: O) From the temperature profile in the liquid; □) from the temperature of the outer surface of the tube.

The results of experiments on the heat transfer beyond the stabilization length are shown in Fig. 5.26. The distribution of the points does not betray any affect on the heat transfer resulting from the additions of sodium.

Johnson, Clabaugh, and Hartnett [18] have studied the heat transfer to mercury flowing in a steel tube (diameter 16.6 mm, length 1220 mm, L/D = 74). The temperatures of the wall of the tube were measured at 8 points, which made it possible to calculate both the local and the mean heat transfer.

Data on the local heat transfer beyond the stabilization length are illustrated in Fig. 5.27, and the results of the heat-transfer measurements in the entrance region are shown in Figs. 5.28, 5.29, and 5.30. Furthermore, a comparison is given here with curves calculated by Deissler, Poppendiek, and Palmer [21]. In Fig. 5.31 are shown also the results of calculations of the heat-transfer coefficient, averaged over the length. A comparison of Figs. 5.27 and 5.31 shows that with large L/D ratios the effect of the entrance region on the average heat transfer is contained within the accuracy of the experiments.

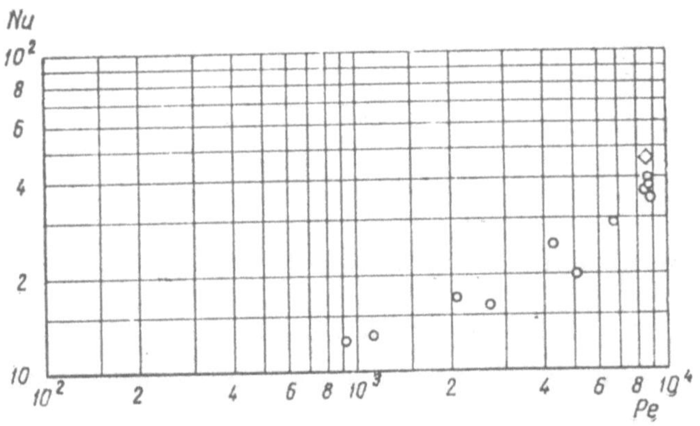

Fig. 5.25. Average heat transfer to mercury flowing in a tube. The temperature of the inner surface of the tube was calculated from a measurement at the outer surface: O) averaging between x/D = 6.3-138, ◇) averaging between x/D = 58-138.

Using this apparatus, the same authors conducted measurements of the heat transfer in the transition and laminar regions of the Re number during the flow of mercury. Figure 5.32 illustrated the results of measurements of the heat transfer beyond the stabilization length, while Fig. 5.33 shows the results of experiments for various points along the entrance region.

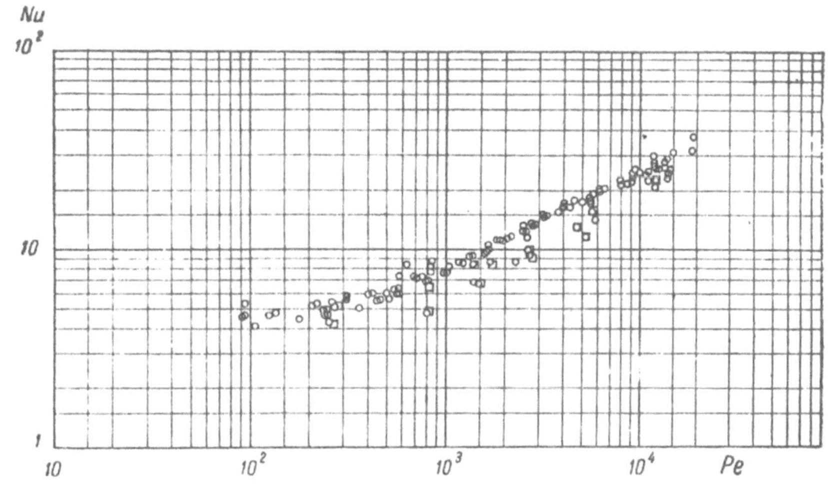

Fig. 5.26. Heat transfer beyond the stabilization length during the flow of mercury in a tube. ○) wetting (addition of sodium; □) nonwetting.

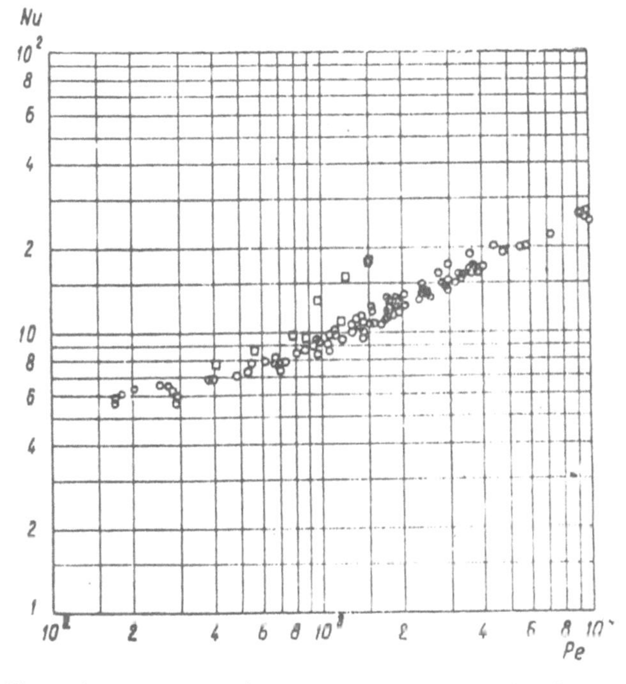

Fig. 5.27. Heat transfer beyond the stabilization length during the flow of mercury in a tube. □) First 30 hours of operation; ○) subsequent tests.

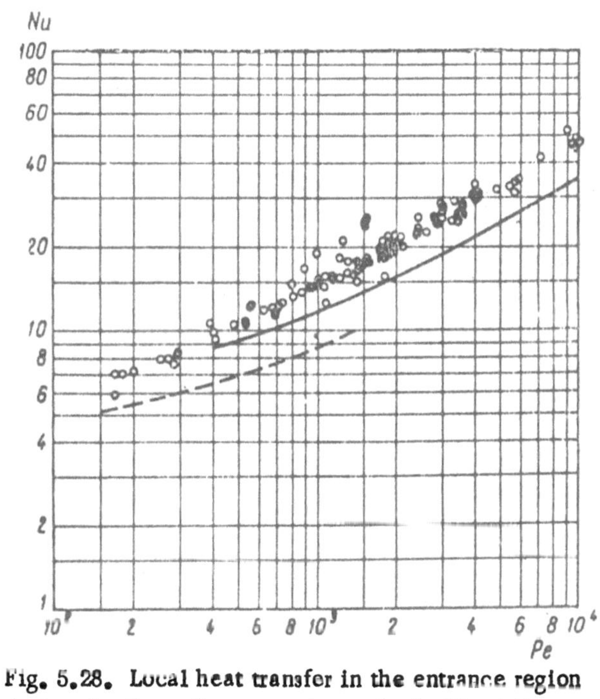

Fig. 5.28. Local heat transfer in the entrance region during the flow of mercury in a tube (x/D = 4.6). ——) Deissler calculation; - - - - -) Poppendiek and Palmer calculation.

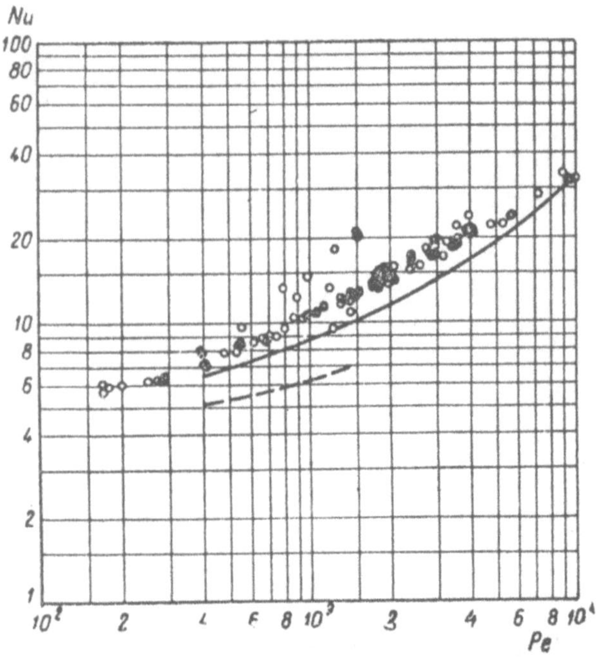

Fig. 5.29. Local heat transfer when x/D = 18.8. The notation is the same as in Fig. 5.28.

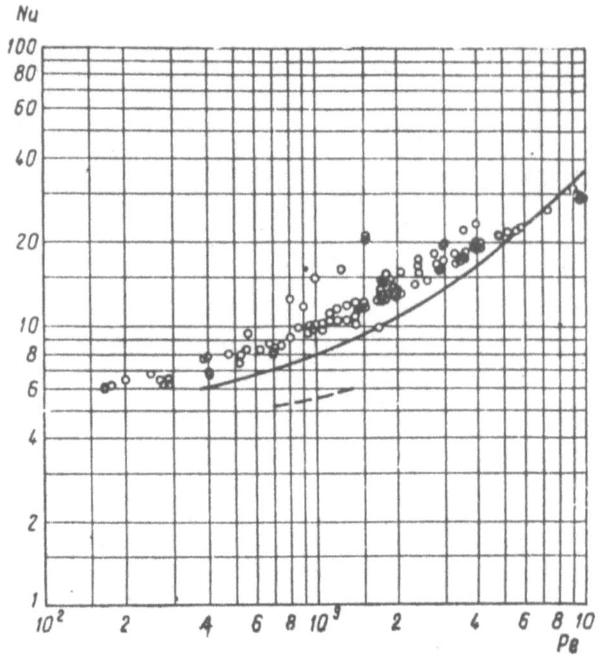

Fig. 5.30. Local heat transfer when x/D = 23. The notation is the same as in Fig. 5.28.

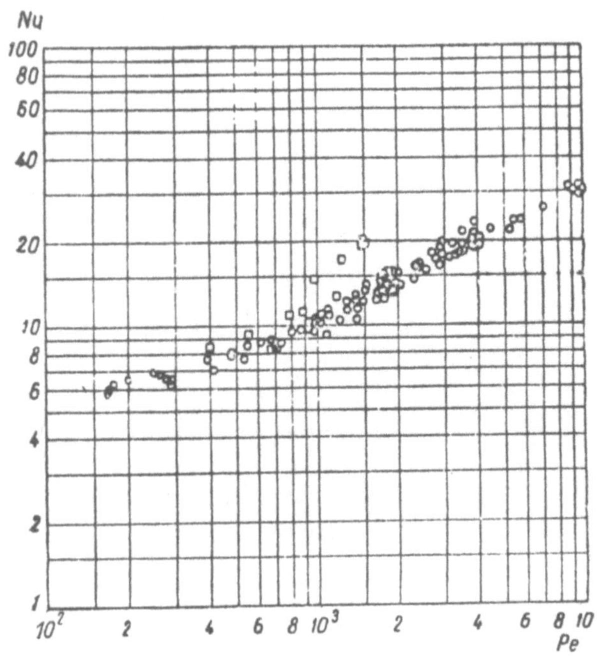

Fig. 5.31. Average heat transfer with mercury flowing in a tube (L/D = 74). □) First 30 hours of operation; ○) subsequent tests.

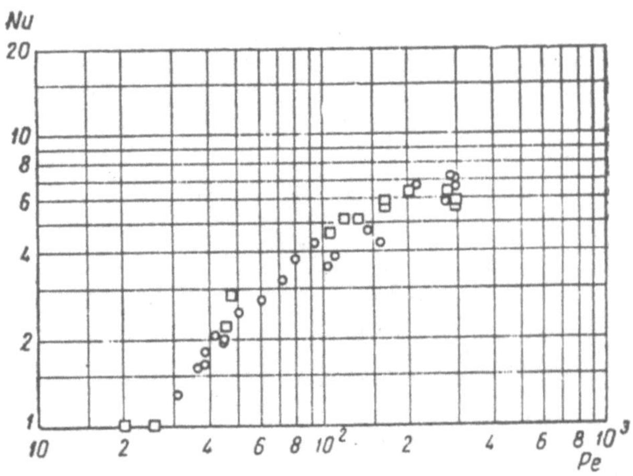

Fig. 5.32. Heat transfer beyond the stabilization length during the flow of mercury and lead-bismuth alloy in a tube (transition and laminar regions of flow). ○) Lead-bismuth; □) mercury.

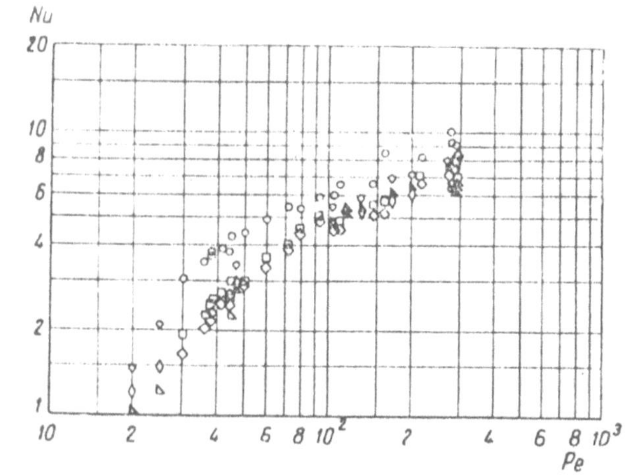

Fig. 5.33. Local heat transfer in the entrance region during the flow of mercury and eutectic lead-bismuth. Lead-bismuth: ○) x/D = 4.6, □) x/D = 13.8, ◇) x/D = 23. Mercury: ▽) x/D = 4.6, ◇) x/D = = 13.8, ◺) x/D = 23.

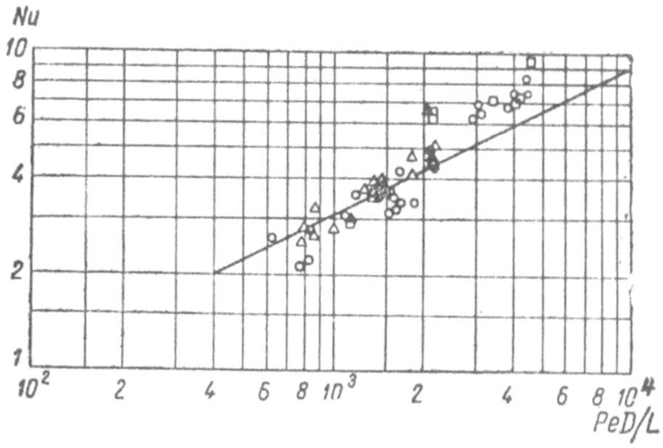

Fig. 5.34. Average heat transfer to mercury flowing in very short channels. Δ, ○) L/D = 1, □) L/D = = 0.5.

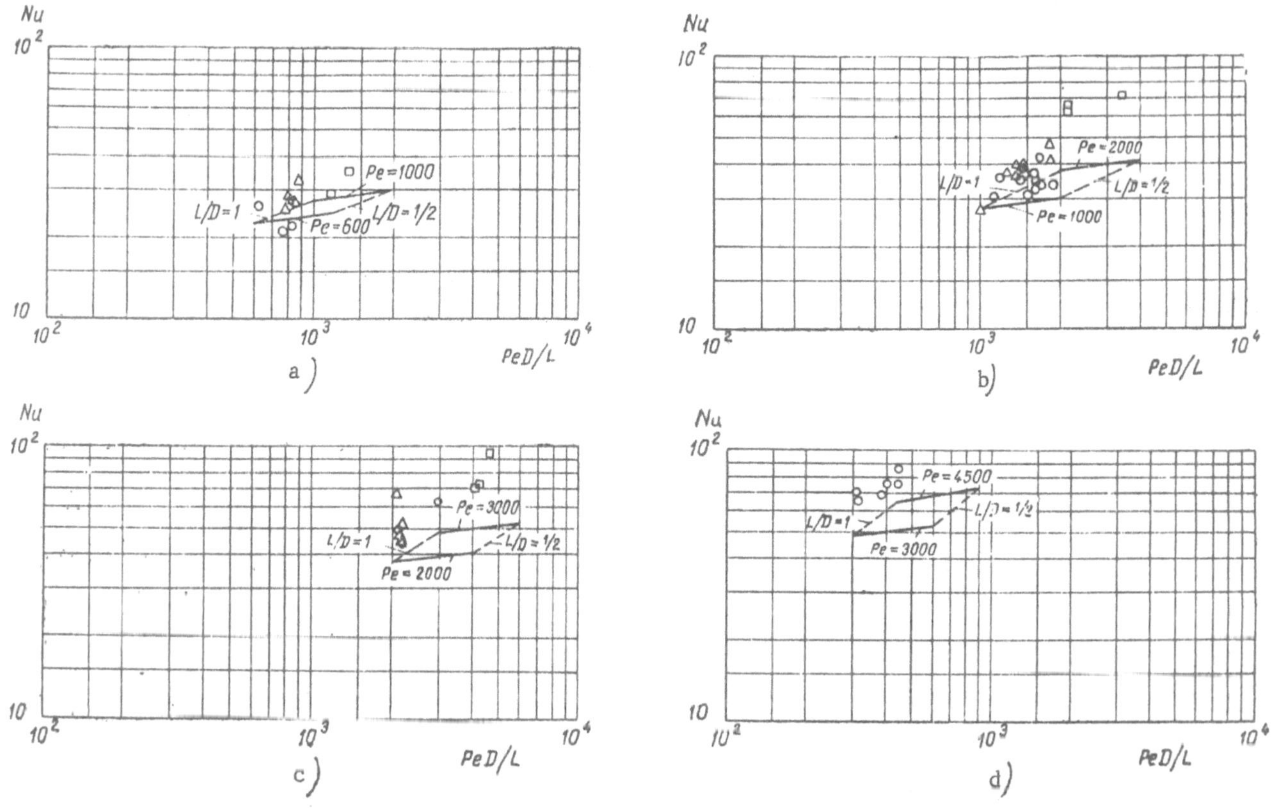

Fig. 5.35. Average heat transfer to mercury flowing in very short channels. Δ, ○) L/D = 1, □) L/D = 0.5; ——— and - - - - - analytic curves [21]. Re number equal to: 600–1000 (a), 1000–2000 (b), 2000–3000 (c), and 3000–4500 (d).

The average heat transfer to mercury flowing in very short channels having diameters of 1.59 and 3.18 mm with L/D equal to 0.5 and 1 were studied by Poppendiek and Harrison [21]. A comparison of the data obtained by these authors for various values of the Pe number and the L/D ratio is given in Fig. 5.34 and Fig. 5.35. The analytic curves of Deissler for the entrance regions are shown there also.

Experimental Data on Heat Transfer To Eutectic Lead-Bismuth

Heat transfer to eutectic lead-bismuth flowing in a tube of diameter 16.6 mm, length 1200 mm was studied by Seban [21]. The wall temperature was measured at 8 points. The method of heating made it possible to obtain conditions approaching t = const.

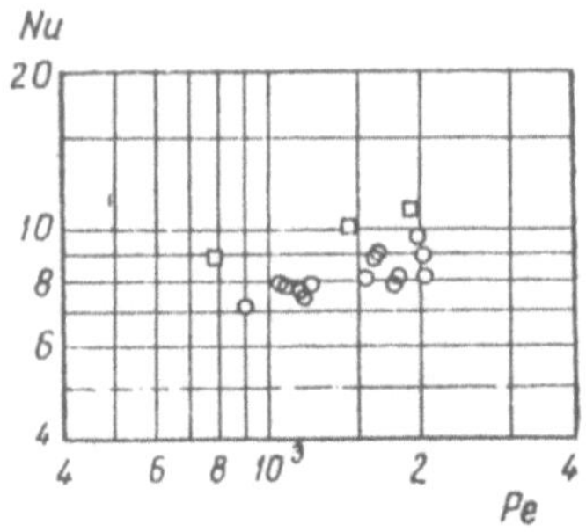

Fig. 5.36. Heat transfer beyond the stabilization length during the flow of eutectic lead-bismuth in a tube. □) Tests with a clean surface during the initial hours of operation; ○) subsequent tests.

Figure 5.36 gives some data on local heat transfer, obtained at the end of the tube beyond the stabilization length. The investigations were made in the Pe number interval 800–2000. In this case the values of Nu varied from 7 to 11. A reduction in the heat-transfer coefficient with the passage of time was observed. In Fig. 5.37 is shown the variation in the ratio of the heat-transfer coefficient at the entrance of the tube to the heat-transfer coefficient beyond the stabilization length. The solid lines correspond to Deissler's data.

The heat transfer in a "figure-eight" heat-exchanger was investigated by Lubarsky [21]. The tests were performed with eutectic lead-bismuth. The length of the tube constituted 1010 mm, the diameter was 10.1 mm, L/D = 100. The inside diameter of the annulus was D_1 = = 12.2 mm, the outside diameter was D_2 = 15.9 mm, the equivalent diameter was $D_e = D_2 - D_1 = 2\delta = 3.7$ mm, and $L/D_e = 322$.

The results of the tests in the tube are shown in Fig. 5.38, those for the annulus are shown in Fig. 5.39.

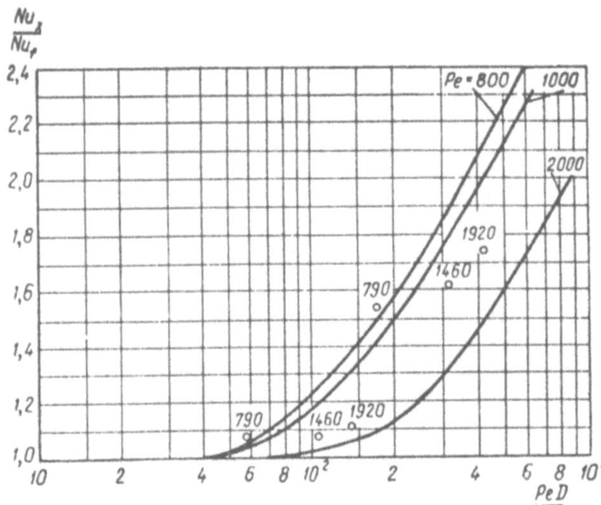

Fig. 5.37. Heat transfer in the entrance region during the flow of eutectic lead-bismuth in a tube. Nu_x is the local heat transfer at a distance x from the entrance, Nu_f is the local heat transfer beyond the stabilization length for various values of Pe, ——) represents the data of Deissler (the numbers on the graph correspond to the values of Pe).

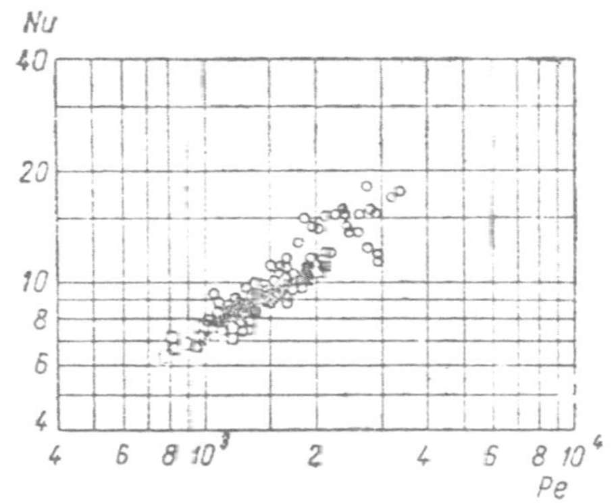

Fig. 5.38. Average heat transfer to eutectic lead-bismuth flowing in a tube. O) Pure eutectic lead-bismuth; □) the same, with an addition of 0.04 % magnesium.

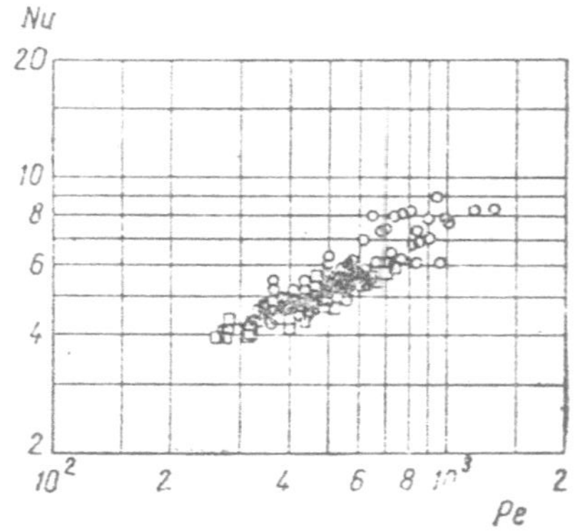

Fig. 5.39. Average heat transfer with eutectic lead-bismuth flowing in an annulus. O) Pure eutectic lead-bismuth; □) the same, with an addition of 0.04% magnesium.

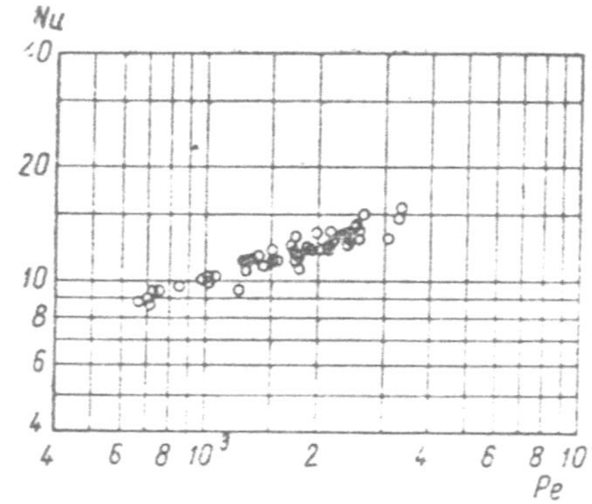

Fig. 5.40. Heat transfer beyond the stabilization length during the flow of eutectic lead-bismuth in a tube.

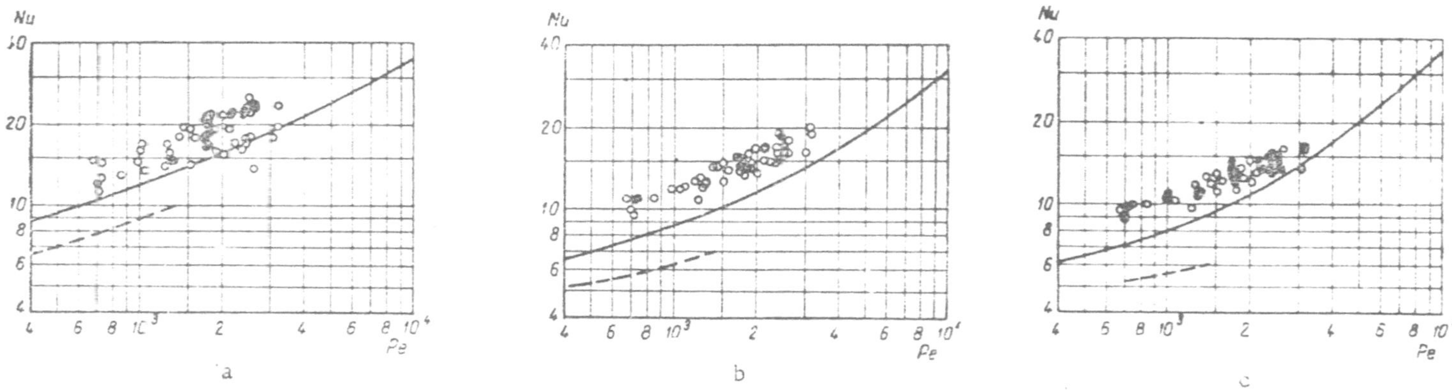

Fig. 5.41. Local heat transfer during the flow of eutectic lead-bismuth in the entrance region of a tube. a) x/D = 4.6, b) x/ D = 13.8, c) x/D = 23; ———) calculation according to Deissler; - - - - -) calculation according to Poppendiek and Palmer.

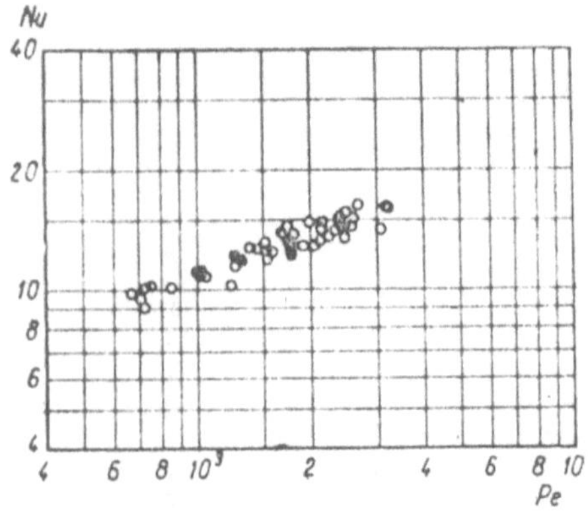

Fig. 5.42. Average heat transfer to eutectic lead-bismuth flowing in a tube.

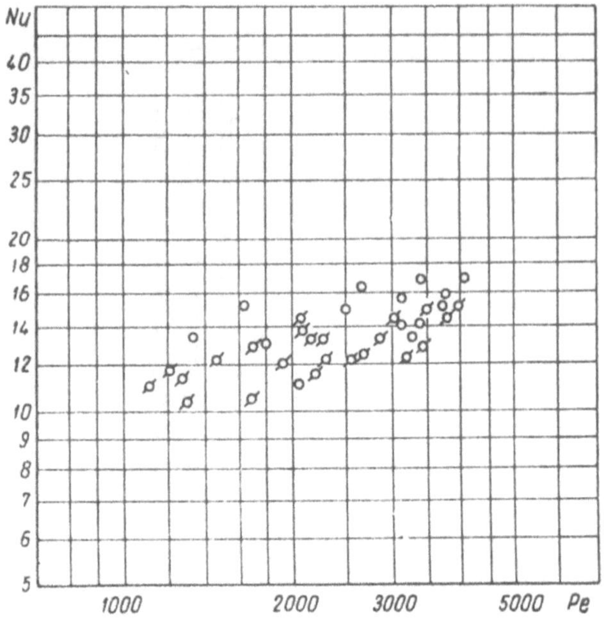

Fig. 5.43. Average heat transfer during the flow of eutectic lead-bismuth in a short tube (L/D ≈ 15) and with the short application of a heating surface. O, ⌀) different series of tests.

Measurements of the heat transfer during the motion of eutectic lead-bismuth in a tube (described above in the section on mercury) were conducted by Johnson, Clabaugh, and Hartnett [18]. The test results for steady-state heat transfer beyond the stabilization length are shown in Fig. 5.40, while the results of heat transfer measurements for the entrance region are shown in Fig. 5.41. Here also is given a comparison with the

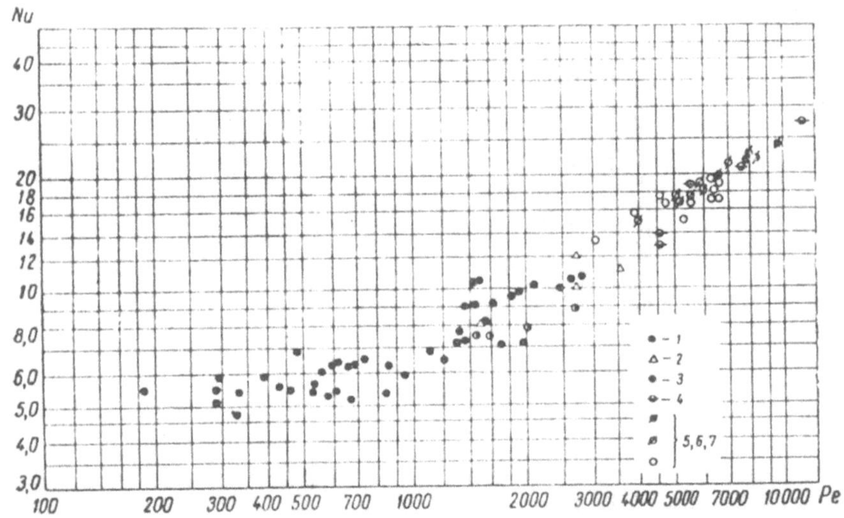

Fig. 5.44. Average heat transfer during the flow of eutectic lead-bismuth in a tube (L/D > 30, flow velocity of 0.15-9 m/sec).

Symbology	Inside diameter of tube, mm	L/D ratio	Remarks
●	9,09	57	
△	9,09	103	
◑	9,09	57	With addition of 0.15% magnesium
-◐-	25,6	38	
● (hatched)	13,4	37	
⌀	9,95	50	
○	5,17	58	

calculations of Deissler, Poppendiek, and Palmer. In these experiments the average heat transfer for all the tubes taken together is also calculated. The results of this calculation for L/D = 74 are shown in Fig. 5.42. The results of measurements of the heat transfer in the laminar and transition regions for the zone of fully stabilized heat transfer and for the entrance region are shown in Fig. 5.32 and Fig. 5.33. The experimental data, those for mercury as well, give sharp reductions in the heat-transfer intensity in the interval Pe = 20 — 300.

An investigation of the average heat transfer for tubes of various lengths (L/D = 15-103) and diameters (9-25 mm) during the flow of eutectic lead-bismuth was made by Borishanskii and Kutateladze [5, 33]. The temperatures of the alloy varied over the interval 210-400° C. Stainless and carbon steel tubes were used. In Fig. 5.43 are shown the test results with a tube with L/D ≈ 15 (similar data were obtained also in experiments with bismuth). Tests with tubes having diameters of 9 mm and L/D equal to 57 and 103 are illustrated in Fig. 5.44. Here also are shown the results of experiments conducted, with Shneiderman participating, for tubes having a diameter of 25.6 mm and L/D = 38. In tests with tubes having diameters of 9.09 and 25.6 mm heating was not maintained (Points 1-4) above $(30-40) \cdot 10^3$ kcal/m^2·hour. In this same figure the results obtained by Ivashchenko for tests with tubes having diameters of 5.17, 9.95, and 13.4 mm and made from steel 20 and Armco iron are given. In these tests the local, rather than the average heat transfer was measured beyond the stabilization length (L/D = 40-50) with a thermal loading of 10^5- $1.3 \cdot 10^6$ kcal/m^2·hour. As evident from the distribution of the points, an agreement between the data obtained in ordinary and stainless steel tubes is observed in a heating interval from $30 \cdot 10^3$ to $1.3 \cdot 10^6$ kcal/m^2·hour (L/D = 30-100). The tests were conducted with the assumption that q = const.

More detailed data on the conditions of these experiments on the heat transfer of a lead-bismuth alloy, and for sodium as well, are given in Table 5.1.

TABLE 5.1
Experimental Heat-Transfer Parameters and Conditions [5, 33]

Inside diameter of tube D, mm	L/D ratio	Material and characteristics of tube	Conditions
		Eutectic lead-bismuth	
8.95	15	Steel EIaIT, industrial tubes	—
9.09	57		$\frac{wD}{a}$ = 170-11,000
9.09	103		
25.0	38	Steel 3, industrial tube	q = 30,000-1,300,000 kcal/m^2·hour
5.17	58	Steel 20 { precision reamed	t_0 = 190-400° C
13.40	37		w = 0.15-9.0 m/sec
9.95	50		
		Sodium	
34	33	Copper, industrial tube	$\frac{wD}{a}$ = 90-1800
35	34		t_0 = 150-350° C w = 0.2-3.5 m/sec q = 50,000-100,000 kcal/m^2·hour

Note. The working section (Fig. 5.44) was made up in the form of a heat wedge consisting of a heat-concentrator having an outside diameter of 100 mm for tubes having diameters of 5.17, 9.95, and 13.4 mm.

The average heat transfer to eutectic lead-bismuth flowing in a carbon steel tube having a diameter of 10.4 mm ($L/D = 34$), under the conditions $q = \text{const}$, were studied by Mikheev, Petrov, Osipov, and Fedynskii [7, 76]. The average temperature of the eutectic alloy in the tests amounted to 270-400° C. The heat flow did not exceed $320 \cdot 10^3$ kcal/m^2·hour. The interval of values for the various moduli were examined as follows: $Re = 20 \cdot 10^3 - 600 \cdot 10^3$, $Pe = 800 - 14{,}000$ and $Pe = 0.018 - 0.025$. The experimental results presented in Fig. 5.45 show the excellent agreement between the experimental points.

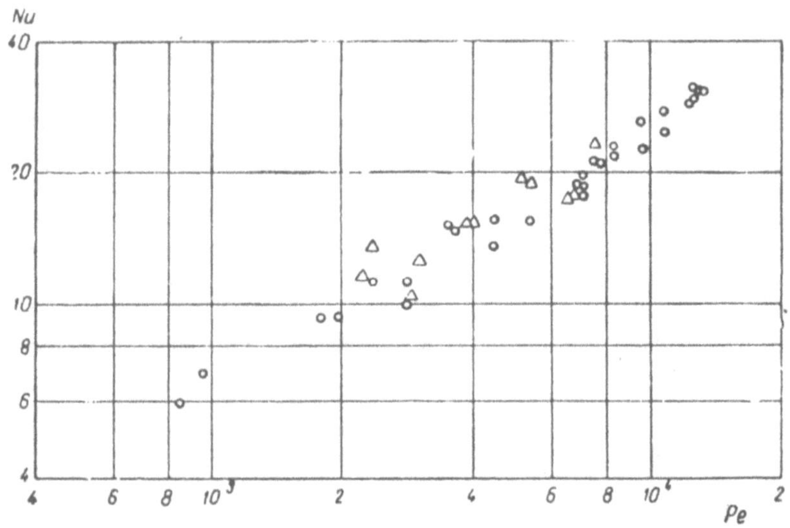

Fig. 5.45. Average heat transfer during the flow of eutectic lead-bismuth and pure bismuth in a horizontal steel tube ($D = 10.4$ mm, $L/D = 34$). △) Bismuth; ○) eutectic lead-bismuth.

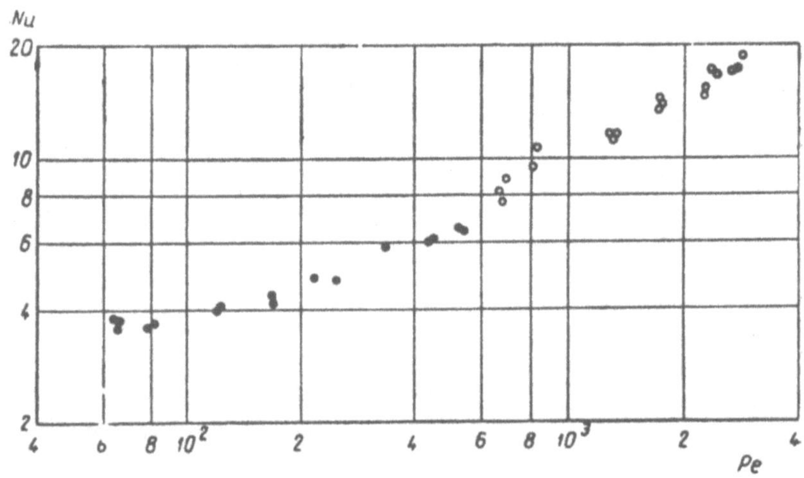

Fig. 5.46. Average heat transfer during the flow of eutectic lead-bismuth in vertical steel tubes. ◉) $D = 10.5$ mm, $L/D = 64$; ○) $D = 5$ mm, $L/D = 34$.

Using similar equipment, these same authors investigated the heat transfer with molten bismuth. The results of these tests are also given in Fig. 5.45. The mean temperature of the bismuth in the tests amounted to 400° C.

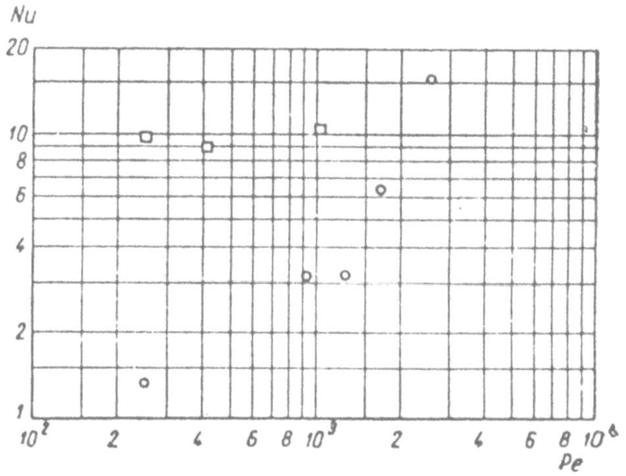

Fig. 5.47. Heat transfer during the flow of eutectic lead-bismuth beyond the stabilization length in a tube. O) Eutectic; □) with addition of 0.05% magnesium (by weight).

A somewhat different distribution of experimental points (Fig. 5.46) was obtained by Voskresenskii and Kondrat'ev in studying the average heat transfer during the flow of eutectic lead-bismuth in tubes made from carbon steel and having diameters of 5 and 10.5 mm (L/D was equal to 34 and 64, respectively). The mean temperature of the alloy in the tests constituted 260-360° C. The specific heat loading did not exceed $160 \cdot 10^3$ kcal/m^2·hour. The values of the Re number were investigated over the interval from 2600 to 150,000.

Measurements of heat transfer for eutectic lead-bismuth in a tube having a diameter of 10 mm and L/D = 63 were carried out by Untermeyer [21] (Fig. 5.47). In contrast with the data cited earlier, the addition of magnesium to the eutectic lead-bismuth increased the heat transfer.

Experimental Data on Heat Transfer To Tin

The average heat transfer to tin flowing in a carbon steel tube having a diameter of 10.4 mm (L/D = 34) was studied by Petrov, Osipov, and Fedynskii [7, 76].

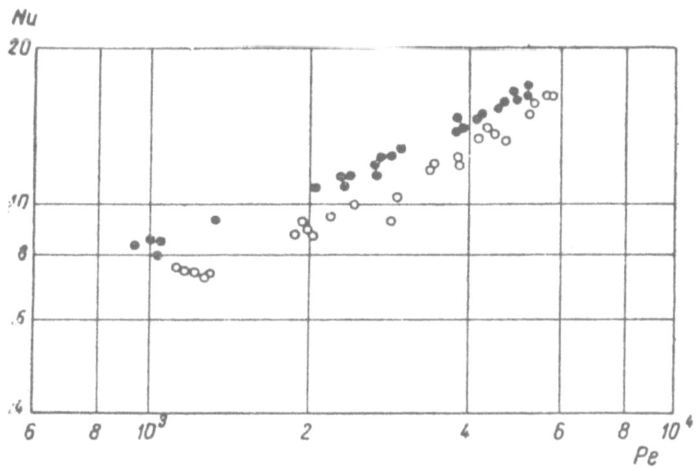

Fig. 5.48. Average heat transfer during the flow of tin in a horizontal steel tube (D = 10.4 mm, L/D = 34). O) Inner surface of the tube machined; ●) inner surface of the tube plated.

In the first series of experiments the inside surface of the tube was machined, while in the second it was plated. In this way two extreme cases were established, namely in which there was very fine oxide layer on the surface and the metal did not wet the surface, and again in which, evidently, thermal contact between the metal of the wall and the flow of molten tin insured definite wetting of the heating surface. The experimental results are shown in Fig. 5.48. It is evident that the experimental points obtained under the conditions of wetting of the heating surface are somewhat higher. Here, the increase in average heat transfer is an approximately constant quantity Δ Nu = 1.5.

Experimental Data on Heat Transfer to Sodium and Eutectic Sodium-Potassium

Tests with an alloy of 52% Na + 48% K were conducted by Lyon in a nickel heat-exchanger (annular). The alloy flowed through the heat-exchanger in both directions. The mass flow of the fluid in the tube and annulus was identical. The dimensions of the inner tube and jacket were varied (the diameters were equal to 11 and 17.8 mm with L/D equal to 111, 98, 76, 159). The dimensions of the annulus were as follows: $\delta = 2.2 - 2.35$ mm, $L = 840 - 1750$ mm.

In the tests the heat transmission was measured, and from this the heat-transfer coefficients for the tube and annulus were determined analytically. The results of the calculations for the tube are shown in Fig. 5.49, while those for the annulus are shown in Fig. 5.50.

Werner, King, and Tidball [21] conducted experiments also with a "figure-eight" heat-exchanger. Sodium-potassium alloys of different compositions flowed in both directions. The alloy flowed in ordinary tubes and in tubes with mixing baffles ("turbulizers") for mixing the fluid before measuring its temperature. The results of the determinations for the tubes are shown in Fig. 5.51, those for the annuli are shown in Fig. 5.52.

The heat transfer to sodium during flow in a copper tube with a diameter of 8.6 mm, $L/D = 46.5$ was studied by Novikov and Khabakhpasheva [22]. In part of the testing the inside surface of the tube was nickel-plated in order to alter the wettability. The flow temperature in the tests was equal to 140-340° C with a flow velocity of 0.8-11 m/sec. The test results are shown in Fig. 5.53. An increase in heat transfer after some definite operating time was observed.

Borishanskii, Kutateladze, Shneiderman, et al. [5, 33] studied the heat transfer to sodium flowing in copper tubes having diameters of 34 and 35 mm with $L/D \approx 34$ (Fig. 5.54). The tests were conducted at a flow temperature of 160-300° C.

The average heat transfer during the flow of sodium in copper tubes (diameters 6.1 and 10.3 mm, L/D equal to 65 and 45, respectively) with $q = \text{const}$ was investigated by Deriugin [76]. Heating in the tests amounted to $5 \cdot 10^4 - 3 \cdot 10^5$ kcal/m²·hour. The mean temperature of the fluid was equal to 100-400° C. The Pr varied from 0.01 to 0.005. The values of Re varied from $10 \cdot 10^3$ to $370 \cdot 10^3$. The test results are shown in Fig. 5.55.

The same author, in collaboration with Mikheev and Fedynskii [76], investigated the heat transfer of sodium in a horizontal copper tube with the application of a heating surface, attaining a delivery of 10^6 kcal/m²·hour. The test results (Fig. 5.56) do not reveal any effects from the magnitude of the heat supply of the heating surface on the intensity of heat transfer.

The same authors (see Fig. 5.57) studied the heat transfer of eutectic sodium-potassium flowing in a red-copper tube (diameter 10.3 mm, $L/D = 45$, mean temperature of the alloy 100-150° C). After completing the test strong wetting of the heating surface was visually observed. The test results are similar to the results obtained in the investigation of the average heat transfer of tin under the conditions of wetting of the heating surface (plated surface).

Hall, Jenkins, and Crofts [23, 47] investigated the heat transfer to sodium and a sodium-potassium alloy in a heat-exchanger having the form of a double annulus with separations from 2 to 3 mm (Fig. 5.58). The sodium temperature varied within the limits 230-280° C, the Re number was greater than 10^4. The temperature of the alloy (22% Na + 78% K) amounted to 100-200° C.

Figure 5.58 shows the sharp drop in heat transfer in the interval $Pe < 100$. During the experiments a marked effect of eccentricity in the annulus on the local heat transfer surrounding the tube was observed. Figure 5.59 shows the experimentally obtained fluctuations in the surrounding temperature, which are expressed as a fraction of the radial temperature fall-off $\Delta t_c / \Delta t_r$, where Δt_c is the over-all temperature drop between the two flows and Δt_r is the drop at a given radius.

The heat transfer with sodium in a tube having a diameter of 16 mm beyond the stabilization length with $x/D \approx 100$ over the interval $Pe \approx 60-200$ was studied by MacDonald and Quittenton [21]. The experimental data are shown in Fig. 5.60. The extensive scatter of the experimental points is evidence of a low precision in the investigations.

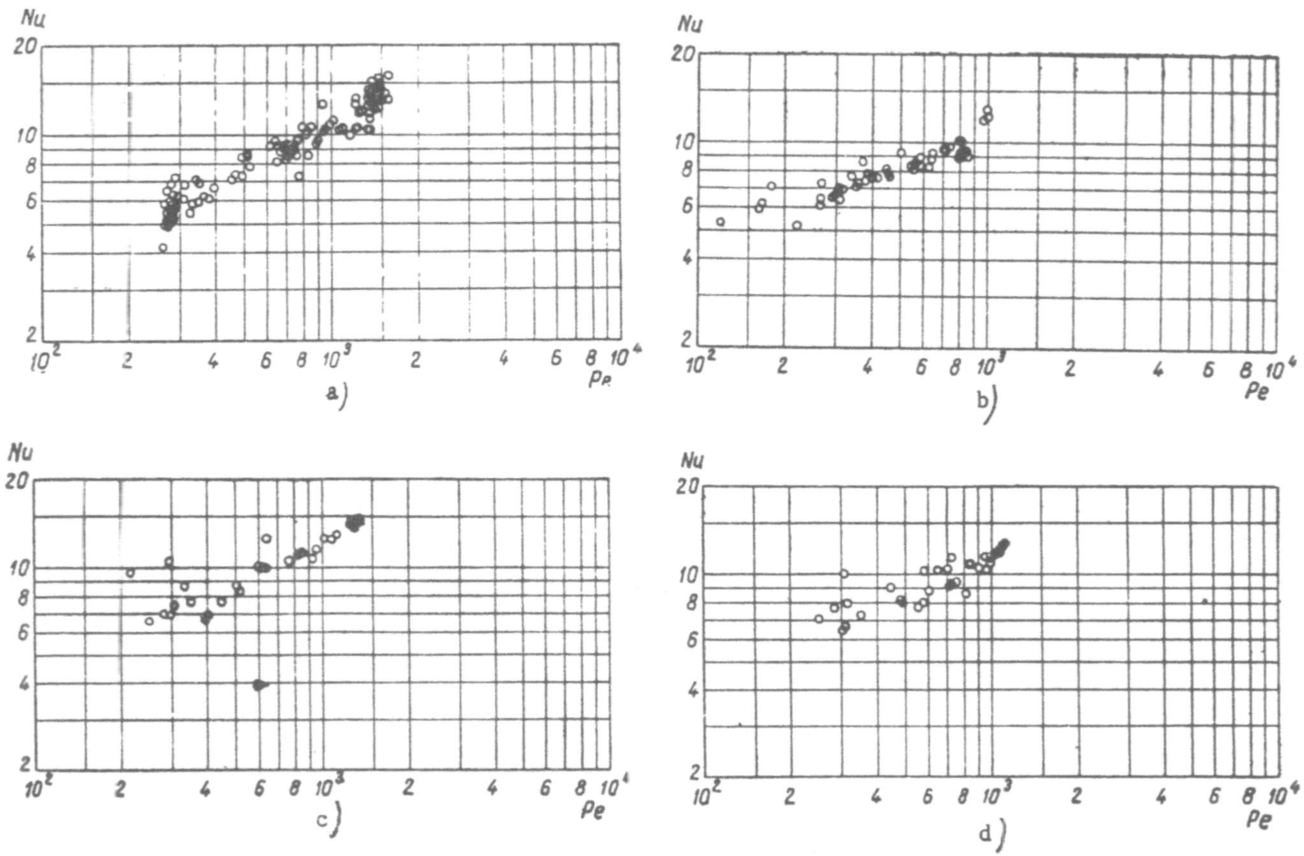

Fig. 5.49. Average heat transfer to a sodium-potassium alloy flowing in a tube. a) L/D = 111, b) L/D = 98, c) L/D = 76, d) L/D = 159.

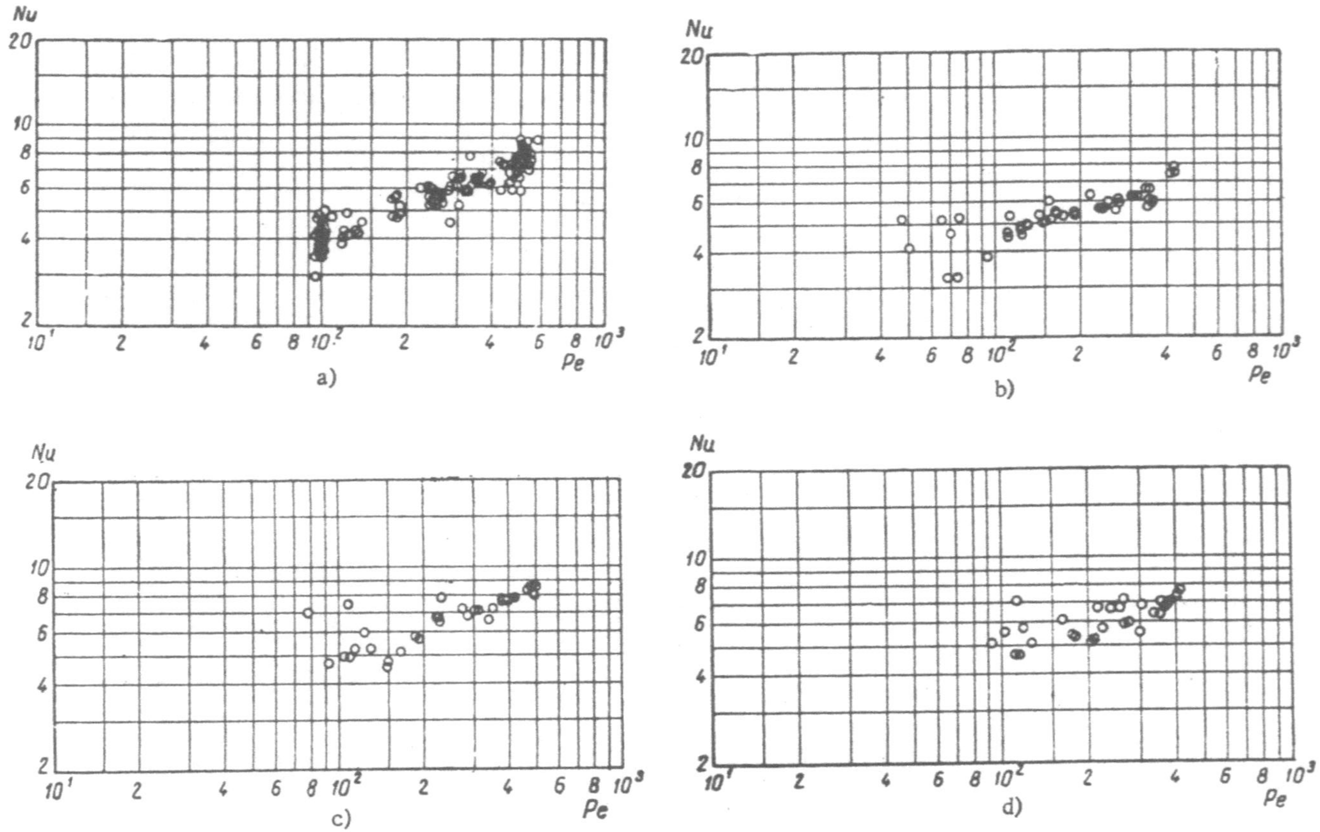

Fig. 5.50. Average heat transfer to a sodium–potassium alloy flowing in an annulus. a) $L/D_e = 223$, $D_2/D_1 = 1.43$, b) $L/D_e = 397$, $D_2/D_1 = 1.23$, c) $L/D_e = 179$, $D_2/D_1 = 1.37$, d) $L/D_e = 375$, $D_2/D_1 = 1.37$.

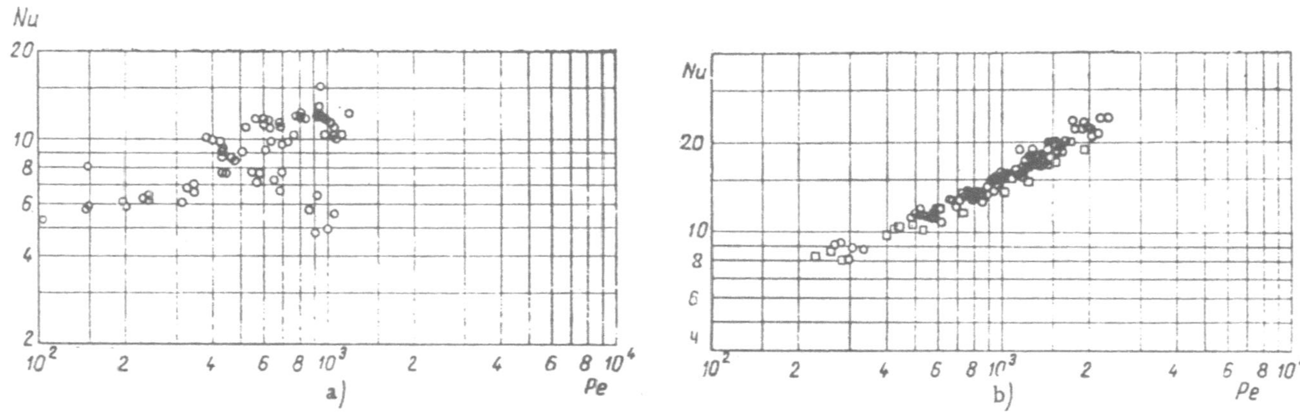

Fig. 5.51. Average heat transfer to a sodium–potassium alloy flowing in a tube. a) Ordinary tube, L/D = 50, b) tube with mixing baffle, L/D = 48; ○) 56% Na + 44% K, □) 23% Na + 77% K.

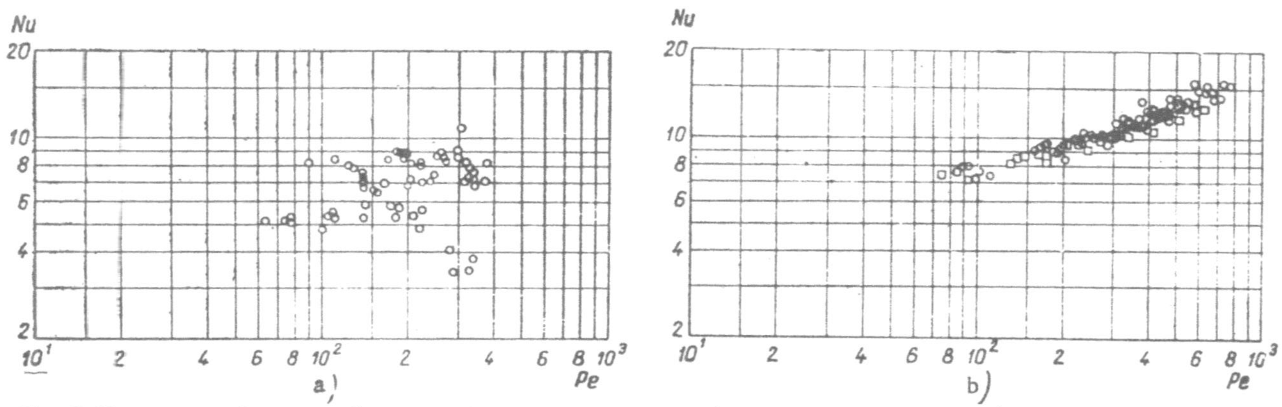

Fig. 5.52. Average heat transfer to a sodium–potassium alloy flowing in an annulus. $L/D_e = 54$, $D_2/D_1 = 1.83$; ○) 56% Na + 44% K; □) 23% Na + 77% K; a) ordinary annulus; b) annulus with mixing baffle.

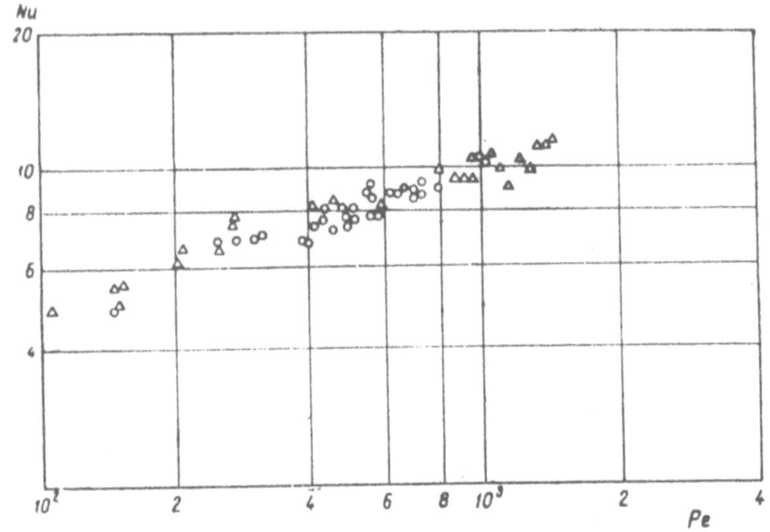

Fig. 5.53. Average heat transfer to sodium flowing in a tube. O) Copper surface, Δ) nickel-plated surface.

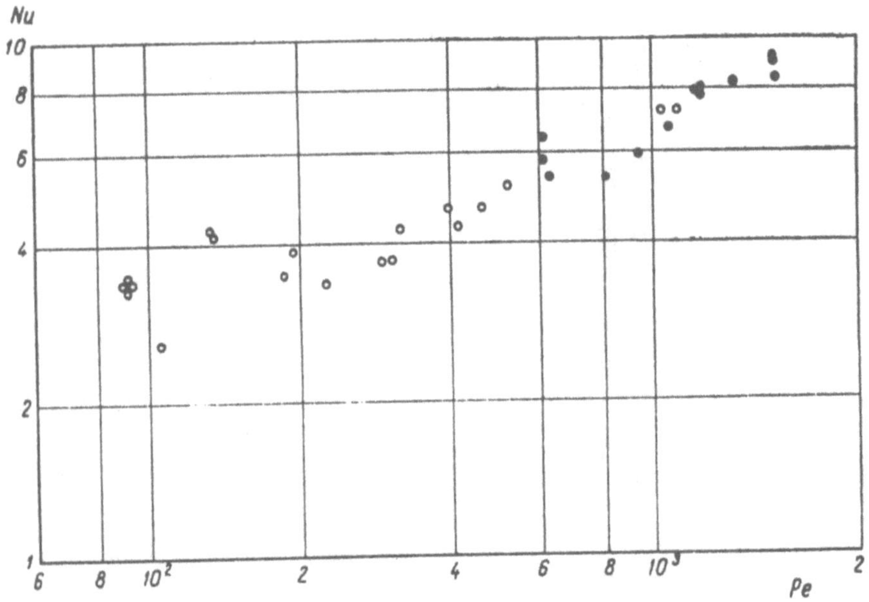

Fig. 5.54. Average heat transfer to sodium flowing in copper tubes. O) D = 34 mm, ●) D = 35 mm.

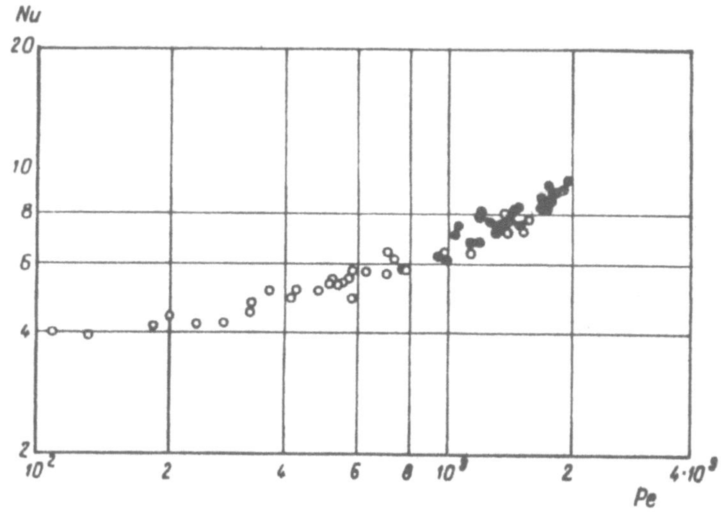

Fig. 5.55. Average heat transfer to sodium flowing in a copper tube. ○) D = 6.1 mm, L/D = 65; ●) D = 10.3 mm, L/D = 45.

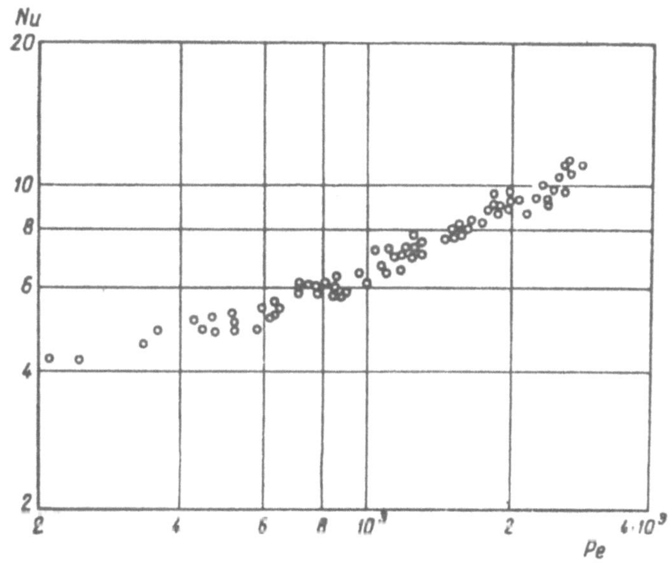

Fig. 5.56. Average heat transfer to sodium flowing in a copper tube. D = 10.3 mm, L/D = 45; $5\cdot10^4 < q < 10^6$.

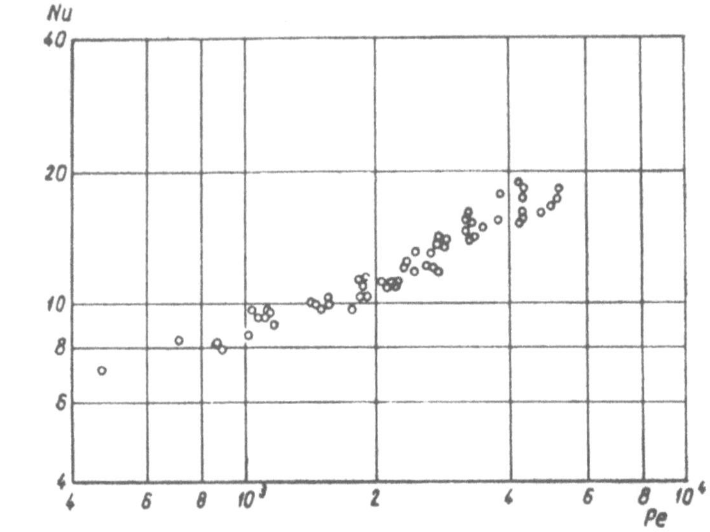

Fig. 5.57. Average heat transfer to a sodium-potassium alloy flowing in a copper tube.

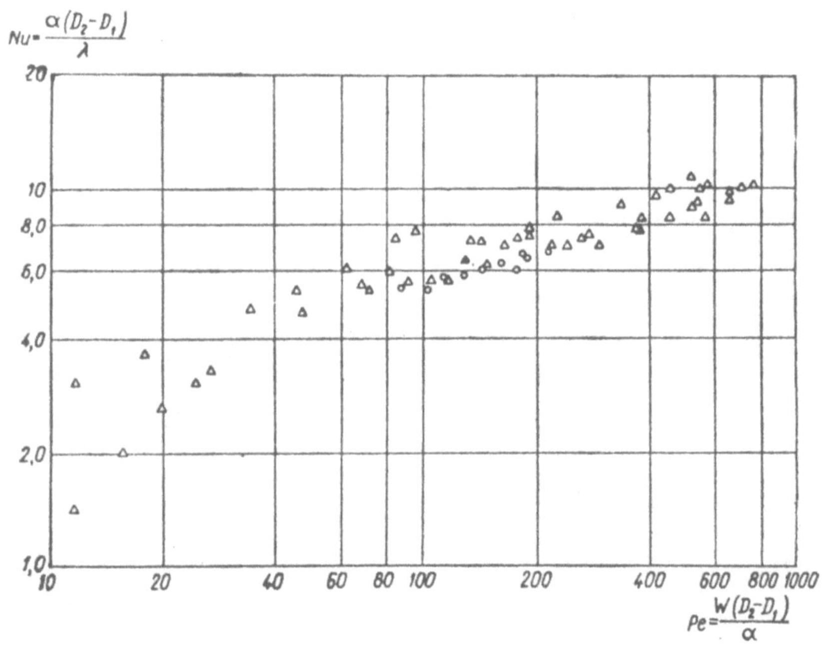

Fig. 5.58. Average heat transfer to a sodium-potassium alloy flowing in annuli. △) Sodium; ○) alloy.

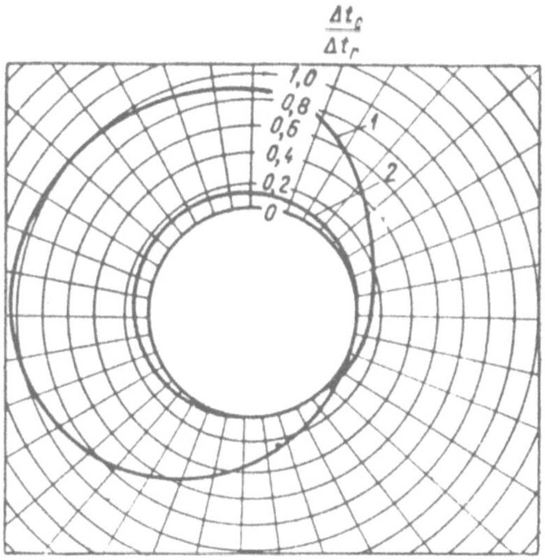

Fig. 5.59. Temperature fluctuations around the outside of a tube with an annular duct. 1) Radial fall-off in tests with a sodium-potassium alloy; 2) radial fall-off in tests with water.

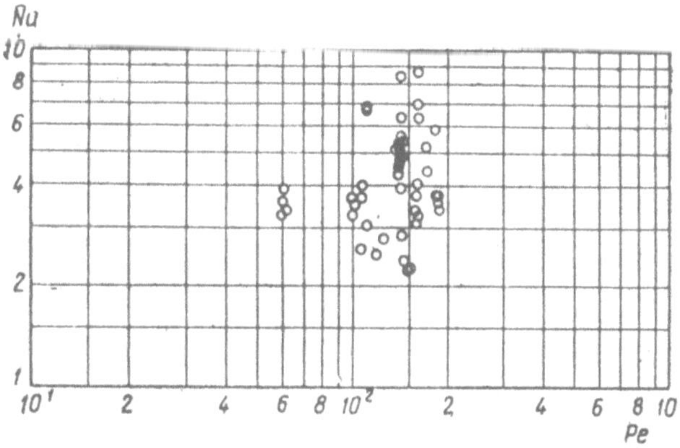

Fig. 5.60. Heat transfer to sodium flowing in a tube.

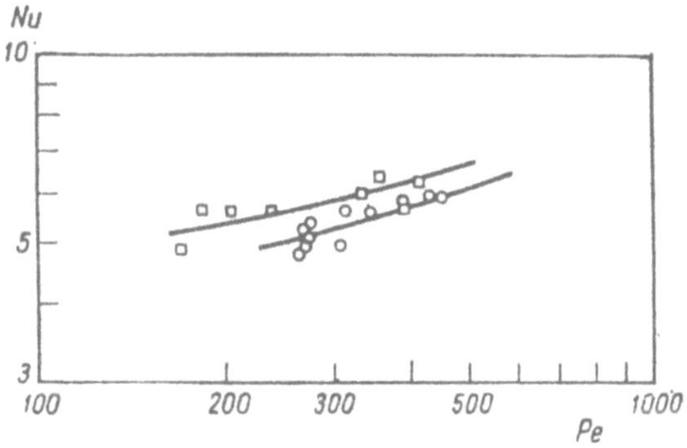

Fig. 5.61. Heat transfer with sodium flowing in a plane channel. ○) Beginning of the test; □) conclusion of the test.

An investigation of the heat transfer in a rectangular channel heat-exchanger with a two-directional flow of a sodium-potassium alloy was conducted by Tidball [24] (Fig. 5.61). The author in this case notes that the heat transfer diminishes somewhat with time.

Comparison of Experimental Data on the Average Heat Transfer In Tubes With L/D > 30

Mercury. The experimental results of different authors studying the average heat transfer with mercury flowing in tubes are compared in Fig. 5.62. In the turbulent regime of flow (Pe > 200) the data of the various authors are spread out considerably. The maximum values of the heat transfer are determined by the experimental data of Isakoff, Drew, and Trefethen, the minimum values are determined from Doody, Younger, Elser, Bailey, et al.

In the transition and laminar regions of flow (Pe < 100) a sharp reduction in the heat transfer down to Nu ~ 1 is observed.

Eutectic lead-bismuth. The averaged experimental results of different authors on the average heat transfer of eutectic lead-bismuth flowing in tubes are cited in Fig. 5.63.

The region Pe > 300 corresponds to the region of fully developed turbulent flow (Re > 10^4). For Peclet numbers Pe < 200 transition and laminar flow occur.

The values of the Nu number, according to Seban's data, are on the average about 7-9 (for Pe = 900-2000), while according to Lubarsky's data they are 6-16 (for Pe = 800-3500).

The experiments of Mikheev, Petrov, Fedynskii, Borishanskii, Kutateladze, Shneiderman, et al. gives values for the Nusselt number Nu = 5-28 (for Pe = 300-12,000).

The experiments of Johnson, Clabaugh, and Hartnett give a somewhat higher value for the heat transfer in the interval Pe = 300-3000. These authors, along with Seban, focussed their attention on the fall-off in the heat transfer in eutectic lead-bismuth as operation of the equipment progressed; they found a reduction amounting to 40-50%. Taken as a whole, it can be stated that the majority of the experimental data in the turbulent flow regime are mutually consistent.

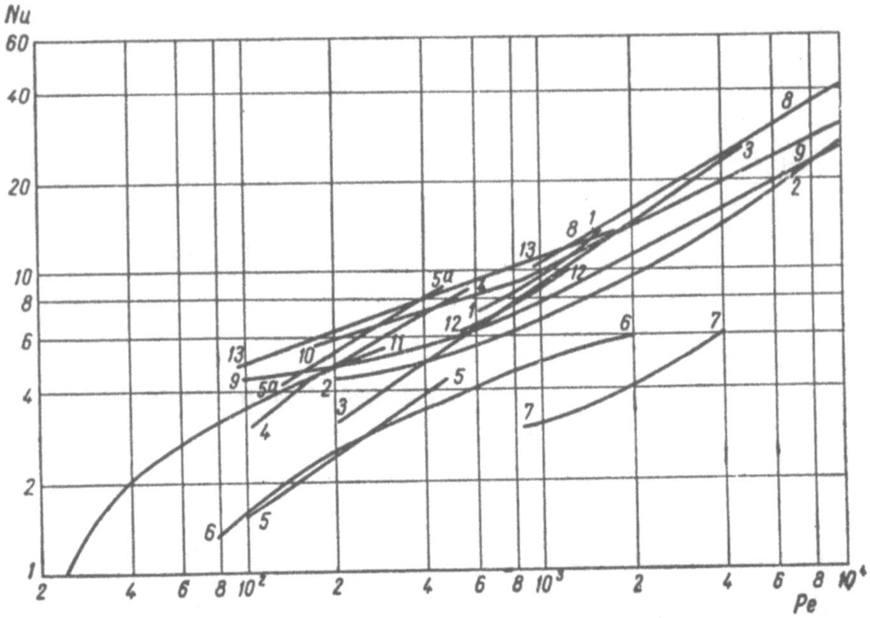

Fig. 5.62. Comparison of the averaged experimental data on heat transfer during the flow of mercury in tubes, L/D > 30. 1) Data of Styrikovich, Sorin, and Semenovker; 2) Voskresenskii, Kalakutskaia, Mikheev, Petrov, and Fedynskii; 3) Korneev; 4) English and Barrett; 5) Doody and Younger; 5a) same, but with sodium added to the mercury; 6) Bailey, Cope, and Watson; 7) Elser; 8) Isakoff and Drew; 9) Stromquist; 10, 11) Johnson, Clabaugh, and Hartnett; 12) Gilliland, Müsser, and Page; 13) Trefethen.

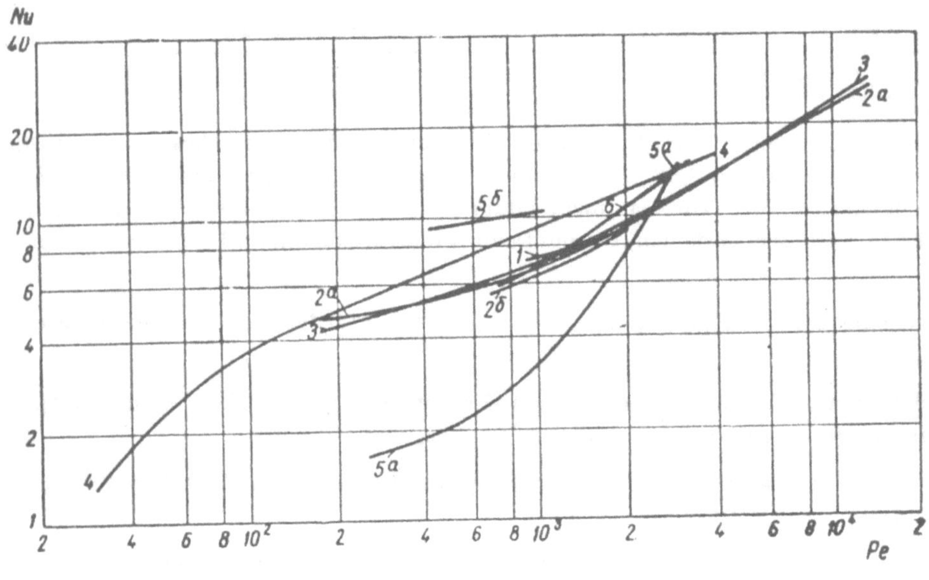

Fig. 5.63. Comparison of the averaged experimental data on heat transfer during the flow of eutectic lead-bismuth in round tubes, L/D > 30. 1) Data of Seban; 2) Borishanskii, Kutateladze, Ivashchenko, and Shneiderman (a) pure eutectic; b) eutectic with an addition of magnesium); 3) Voskresenskii, Kondrat'ev, Mikheev, Petrov, and Fedynskii; 4) Johnson, Clabaugh, and Hartnett; 5) Untermeyer (a) pure eutectic; b) with an addition of magnesium); 6) Lubarsky.

In the interval Pe = 100-800 there are only three sets of experimental data. The data of Borishanskii and Kutateladze are almost the same as those of Mikheev, Petrov, and Fedynskii, while Johnson's data is in agreement with these for Pe ≈ 100-200 and are distributed somewhat higher for larger Pe numbers. In the region of small Pe numbers the heat transfer falls off in intensity from the value Nu = 3.5 (for Pe ≈ 100) to Nu ≈ 1 (for Pe ≈ 30). The interval of Pe numbers from 30 to 60 is the transition region (Re ≈ 1500-3000).

Sodium. Averaged data on the heat transfer with metallic sodium flowing in copper tubes are compared in Fig. 5.64. The interval Pe = 100-2500 was investigated.

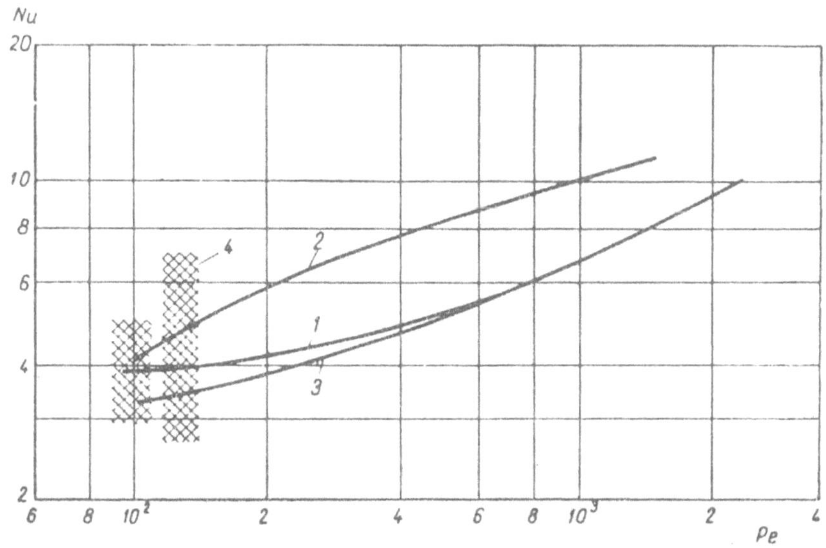

Fig. 5.64. Comparison of the averaged experimental data on heat transfer during the flow of metallic sodium in round tubes, L/D > 30.
1) Data of Deriugin, Mikheev, and Fedynskii; 2) Novikov and Khabakhpasheva; 3) Borishanskii, Kutateladze, and Shneiderman; 4) MacDonald and Quittenton (striated region).

The experimental results of Deriugin and Fedynskii almost agree with those of Borishanskii, Kutateladze, and Shneiderman. The test data of Novikov and Khabakhpasheva are distributed higher, but approach the data of the preceding authors somewhat as the values of the Pe number increase. In these last experiments an increase in the heat transfer as the operation of the equipment progressed was noted.

Eutectic sodium-potassium. The averaged results of the various authors investigating the flow of eutectic sodium-potassium in round tubes are compared in Fig. 5.65. For values of the Peclet number Pe ≈ 1000 the experiments of Lyon give Nu = 11-14, those of Werner, King, and Tidball give Nu = 10-15, and those of Deriugin, Mikheev, and Fedynskii give Nu ≈ 8.5.

Thus, it can be stated that a considerable spread exists in the experimental data on the average heat transfer during the flow of eutectic sodium-potassium in tubes.

Influence of additives. A comparison of the investigated papers does not give a clear representation of the effect of surface-active substances on heat transfer to liquid metals. Tests with magnesium-mercury amalgams, performed by Korneev [14], and Stromquist's tests [21] with additions of sodium to mercury do not disclose any variation in the heat transfer. However, the tests of Doody and Younger [16], conducted with the same Pe numbers, demonstrate a contrary result.

The tests of Lubarsky [21] and Borishanskii and Kutateladze [5, 33] with additions of magnesium to eutectic lead-bismuth did not lead to significant improvements in the heat transfer. At the same time, however, Untermeyer [21] in a similar case observed a considerable increase in the heat transfer. A comparison of the experimental data, for the purpose of exhibiting the effect of the material of the tube walls on the heat transfer, allows us to make the statement that qualitatively the heat transfer, apparently, is higher when the metal wets the heating surface. However, a quantitative determination of this effect is not possible at present.

In order to calculate the heat transfer with liquid metals during flow in industrial tubes we can recommend the equations presented in [5, 7, 33, and 76]:

$$Nu = 3{,}3 + 0{,}014\, Pe^{0{,}8}, \tag{5.20}$$

$$Nu = 5 + 0{,}0021\, Pe. \tag{5.21}$$

Equation (5.20) is recommended for application in the interval 200 < Pe < 20,000, while the somewhat simpler equation (5.21) will serve for the interval 300 < Pe < 15,000. In both cases Re > 10^4.

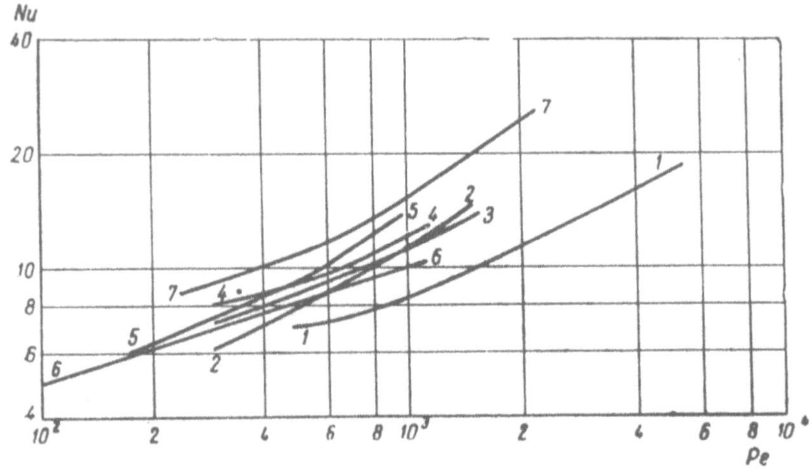

Fig. 5.65. Comparison of the experimental data on heat transfer during the flow of a sodium-potassium alloy in round tubes, L/D > 30. 1) Data of Derlugin, Mikheev, and Fedynskii; 2-5) Lyon; 6, 7) Werner, King, and Tidball.

In the region of small Pe numbers a tentative calculation can be made from the equation [5, 33, and 42]:

$$Nu = 0{,}7\, Pe^{1/3}. \tag{5.22}$$

The possible application of this equation is limited to the interval 20 < Pe < 300.

For calculating the heat transfer when a high degree of purity is maintained in the operation and there is wetting by a molten metal, an equation given in [76] is recommended:

$$Nu = 4{,}8 + 0{,}014\, Pe^{0{,}8}. \tag{5.23}$$

The range of applicability of this equation is the same as that for equation (5.20).

Equations (5.20) - (5.23) are recommended for any alkali molten metals and their alloys.

For sodium, under the condition of reliable wettability of the heating surface and assuming maintenance of the proper purity during the operation, an equation cited in [22] is recommended:

$$Nu = 5{,}9 + 0{,}015\, Pe^{0{,}8}, \tag{5.24}$$

which is valid for 100 < Pe < 1400.*

All of these equations make it possible to calculate the heat transfer, averaged over the length of the tube, for L/D > 30.

* It should be noted that in applying (5.23), as opposed to a calculation with (5.24), a certain safety factor is provided for.

Heat transfer in the entrance region, as well as in short tubes, has been examined in a number of papers, and a comparison of the experimental data is made in papers by Kondrat'ev [75] and Lubarsky and Kaufman [21].

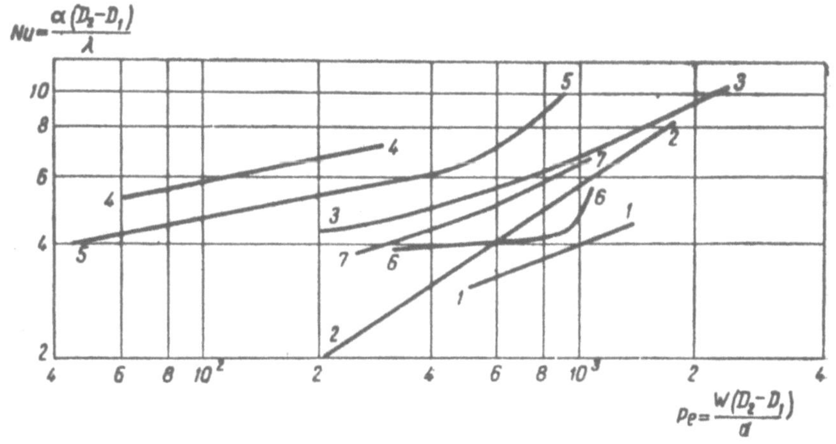

Fig. 5.66. Comparison of the experimental data on heat transfer during the flow of mercury and a lead-bismuth alloy in annuli. Mercury: 1) data of Styrikovich, Sorin, and Semenovker ($D_2/D_1 = 1.71$); 2) Korneev ($D_2/D_1 = 1.35$); 3) Fedynskii ($D_2/D_1 = 1.55$); 4) Trefethen ($D_2/D_1 = 1.4$); 5) Trefethen ($D_2/D_1 = 1.75-2.31$); 6) Sineath (rectangular annulus). Eutectic lead-bismuth: 7) data of Lubarsky ($D_2/D_1 = 1.25$).

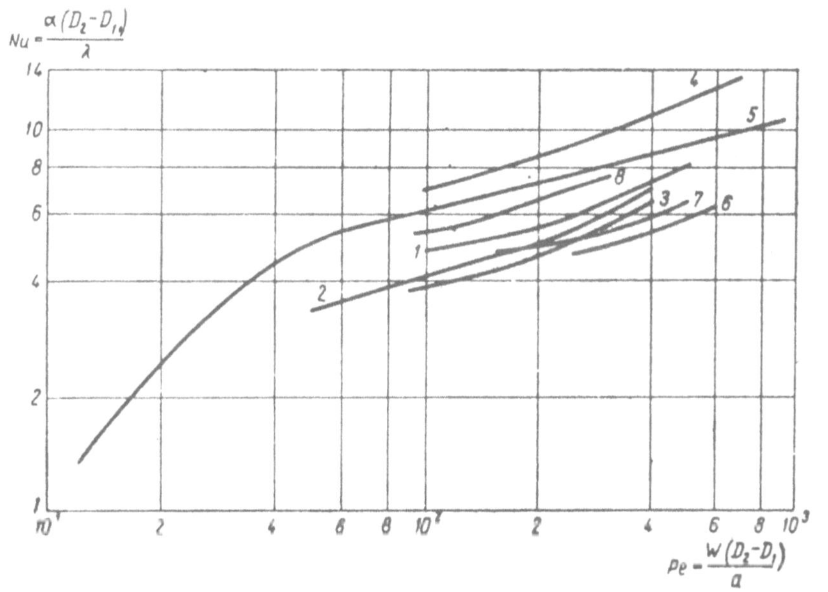

Fig. 5.67. Comparison of the experimental data for the flow of sodium and a sodium-potassium alloy in annuli. Sodium-potassium: 1) data of Lyon ($D_2/D_1 = 1.37, 1.43$); 2) Lyon ($D_2/D_1 = 1.23$); 3) Werner, King, and Tidball ($D_2/D_1 = 1.83$-steel); 4) the same, but for nickel; 5) Hall, Crofts, and Jenkins ($D_2/D_1 = 1.3$); 6) Tidball (rectangular annulus, beginning of operation); 7) the same, but at the end of the operation. Sodium: 8) Hall, Crofts, and Jenkins.

From the comparison of the experimental data of the various researchers it can be assumed that for L/D > > 30 the effect of the entrance region is not very pronounced in the average heat transfer. A calculation of the heat transfer in the entrance region and in small tubes can be made from the equations presented above, if we introduce the correction factor [7]:

$$\varepsilon_l = 1{,}72\left(\frac{D}{L}\right)^{0,16}. \qquad (5.25)$$

Comparison of Experimental Data on the Heat Transfer in Annuli

In Fig. 5.66, the experimental data on heat transfer during the flow of mercury and eutectic lead-bismuth in circular and rectangular annuli are compared. In Fig. 5.67, a similar comparison is made for experimental data obtained in rectangular and circular annuli with a flow of sodium and a sodium-potassium alloy.

Experiments with the heavy metals (mercury, eutectic lead-bismuth) took in an over-all interval Pe = = 45-2200 while those with the alkali metals (sodium, sodium-potassium alloy) were made over the interval of Peclet numbers from 12-1000.

A comparison of all these data make it possible, to a first approximation, to make calculations from the equations for the heat transfer involved in the flow of liquid metals in round tubes, introducing the hydraulic equivalent diameter $D_e = D_2 - D_1 = 2\delta$, where δ is the annular width.

Chapter 6

HEAT TRANSFER DURING LONGITUDINAL FLOW AROUND A PLATE

Some Features of Heat Transfer When A Fluid Having Pr ≪ 1 Flows Around An Obstacle

It is a well-known fact that for the process of heat transfer between a body and a fluid, those phenomena which take place in the immediate vicinity of the solid surface against which the fluid is moving are the most essential ones. The flow region in which a change occurs in the relative streaming velocity from zero to a quantity which is almost the same as the velocity of the undisturbed flow is called the hydrodynamic boundary layer. The flow region in which there is a similar change in temperature is called the thermal boundary layer.

In the general theory of convection heat exchange it is proven that the thickness of the hydrodynamic and thermal boundary layers exactly coincide only in a medium which has Pr = 1, and which flows without variation in pressure when similar boundary conditions are specified with respect to the flow velocities and temperatures.

In all the remaining cases these thicknesses are not equal to one another and their relationship depends primarily on the relationship between the kinematic viscosity and diffusivity of the medium, i.e., on the Pr number.

Since for $a > \nu$ the region of thermal perturbations is propagated further into the flow than is the region of hydrodynamic perturbations, in liquid metals the thermal boundary layer is thicker than the hydrodynamic boundary layer.

In this case the law which determines the progress of the dynamic boundary layer in the liquid metal naturally remains the same as in other liquids subjected to Newton's law of friction forces. The heat-transfer laws in liquid metals, however, are substantially altered.

The presentation of this problem following is based on calculations carried out by Kutateladze and Fedorovich.

The heat-exchange coefficient for flow around an infinite plane obstacle by a flow of some incompressible fluid is determined by the following integral equation:

$$\frac{\alpha_x}{c\gamma w_0} = \frac{d\delta_T^{**}}{dx} + \left[\frac{1}{w_0}\frac{dw_0}{dx} + \frac{1}{t_W - t_0}\frac{d(t_W - t_0)}{dx}\right] \times \delta_T^{**} + \frac{1}{t_W - t_0}\frac{dt_0}{dx}\int_0^{\delta_T} \frac{w_x}{w_0}\,dy, \tag{6.1}$$

where $\delta_T^{**} = \int_0^{\delta_T} \frac{w_x}{w_0}\left(1 - \frac{t_W - t}{t_W - t_0}\right) dy$ is the so-called heat-content loss interval, δ_T is the thickness of the thermal boundary layer, α_x is the local value of the heat-transfer coefficient, t_s is the temperature of the heat-exchanging surface, t_0 is the temperature of the unperturbed flow, t is the temperature at a point having the coordinates (x, y), w_0 is the velocity of the unperturbed flow, w_x is the component of the velocity vector having the coordinate x, x is a coordinate directed along the surface of the body in the direction of the vector w_0, y is a coordinate directed along the normal to the surface of the body at a given point.

For a plate having a constant temperature t_s and immersed in an unbounded fluid flow, with w_0 = const and t_0 = const, equation (6.1) can be written as follows:

$$\frac{\alpha x}{c\gamma w_0} = \frac{d}{dx}\int_0^{\delta_T} \frac{w_x}{w_0}\left(1 - \frac{t_W - t}{t_W - t_0}\right) dy. \qquad (6.2)$$

For the solution of this equation it is necessary to know the velocity and temperature profiles in the boundary layer.

Since in liquid metals $\delta_T > \delta$ (δ is the thickness of the hydrodynamic boundary layer), in the above equations the velocity profile appears to be broken up into two parts. In the interval $0 < y < \delta$ the velocity varies according to the ordinary law for a hydrodynamic boundary layer of an incompressible fluid. In the interval $y > \delta$ the velocity can be considered constant and equal to the velocity of the undisturbed flow w_0. Here the problem of the effect of turbulence on the temperature profile outside the hydrodynamic boundary layer becomes essential. This problem does not arise in considering nonmetallic liquids, since in these we have $\delta \gtrsim \delta_T$.

Laminar Boundary Layer Flowing Around A Plate

Let us assume that the free turbulence of a moving stream is very small and its effect on the heat transfer in the region $\delta < y < \delta_T$ can be neglected.

Then the temperature distribution in the thermal boundary layer can be described by the familiar interpolation expression

$$\theta = \frac{t_W - t}{t_W - t_0} = 2\left(\frac{y}{\delta_T}\right) - 2\left(\frac{y}{\delta_T}\right)^3 + \left(\frac{y}{\delta_T}\right)^4. \qquad (6.3)$$

The coefficients of this equation are chosen in such a way that the boundary conditions below are satisfied:

$$y = 0, \quad t = t_W, \quad \frac{\partial^2 t}{\partial y^2} = 0;$$

$$y = \delta_T, \quad t = t_0, \quad \frac{\partial t}{\partial y} = 0, \quad \frac{\partial^2 t}{\partial y^2} = 0.$$

Similarly, the velocity profile inside the hydrodynamic boundary layer is described by the expression

$$\frac{w_x}{w_0} = 2\left(\frac{y}{\delta}\right) - 2\left(\frac{y}{\delta}\right)^3 + \left(\frac{y}{\delta}\right)^4. \qquad (6.4)$$

It should be emphasized that the distribution of the dimensionless velocity w_x/w_0 and the dimensionless temperature θ in this case are mutually convergent due to the similarity of the boundary layers. However, they are not similar to one another, since $\delta \neq \delta_T$.

Substituting the expressions (6.3) and (6.4) into equation (6.2) and bearing in mind that when $y > \delta$, $w = w_0$, we obtain

$$\frac{\alpha x}{c\gamma w_0} = \frac{d}{dx}\int_0^{\delta} \frac{w_x}{w_0}(1-\theta)\, dy + \frac{d}{dx}\int_{\delta}^{\delta_T}(1-\theta)\, dy =$$

$$= \frac{d}{dx}\int_0^{\delta}\left[2\frac{y}{\delta} - 2\left(\frac{y}{\delta}\right)^3 + \left(\frac{y}{\delta}\right)^4\right]\left[1 - 2\frac{y}{\delta_T} + 2\left(\frac{y}{\delta_T}\right)^3 -\right.$$

$$\left. - \left(\frac{y}{\delta_T}\right)^4\right] dy + \frac{d}{dx}\int_{\delta}^{\delta_T}\left[1 - 2\frac{y}{\delta_T} + 2\left(\frac{y}{\delta_T}\right)^3 - \left(\frac{y}{\delta_T}\right)^4\right] dy. \qquad (6.5)$$

On the other hand,

$$\alpha_x = (t_W - t_0) = -\lambda\left(\frac{\partial t}{\partial y}\right)_{y=0} = \frac{2\lambda}{\delta_T}(t_W - t_0), \tag{6.6}$$

whence

$$\frac{\alpha_x}{c\gamma w_0} = \frac{2a}{w_0 \delta_T}. \tag{6.7}$$

Solving equations (6.5) and (6.7) simultaneously, we obtain

$$\frac{2a}{w_0\delta_T} = \frac{d}{dx}\left\{0{,}300\delta_T\left[1 - \frac{\delta}{\delta_T} + 0{,}410\left(\frac{\delta}{\delta_T}\right)^2 - \right.\right.$$
$$\left.\left. - 0{,}0735\left(\frac{\delta}{\delta_T}\right)^4 - 0{,}146\left(\frac{\delta}{\delta_T}\right)^5\right]\right\}. \tag{6.8}$$

The solution of the problem of heat exchange in the laminar boundary layer of nonmetallic fluids shows that the ratio δ/δ_T can with a high degree of accuracy be considered a function of the physical properties of the fluid only. To a good approximation it can be considered that

$$\frac{\delta}{\delta_T} \approx Pr^{1/3}. \tag{6.9}$$

If, using this equation, we estimate the values of the terms in equation (6.8), then it becomes clear that even for Pr = 0.1 in the equation it is possible to discard all quantities containing the ratio δ/δ_T to powers greater than one.

Integrating equation (6.8) after simplification in this manner, we find that

$$\delta_T = 3{,}65\sqrt{\frac{ax}{\left(1 - \frac{\delta}{\delta_T}\right) w_0}}. \tag{6.10}$$

The constant of integration is equal to zero, since for x = 0, $\delta_T = 0$.

Substituting this value of δ_T into equation (6.7), we obtain

$$\alpha_x = 0{,}82\sqrt{\frac{\left(1 - \frac{\delta}{\delta_T}\right) w_0}{ax}}. \tag{6.11}$$

The thickness of the laminar hydrodynamic boundary layer at the plate is determined by the familiar equation

$$\delta = 5{,}83\sqrt{\frac{\nu x}{w_0}}. \tag{6.12}$$

Thus, in the present approximation

$$\frac{\delta}{\delta_T} = \frac{2Pr^{1/2}}{Pr^{1/2} + \sqrt{Pr + 1{,}56}}. \tag{6.13}$$

From this it follows that when Pr → 0 the quantity $\delta/\delta_T \to 1.16\ Pr^{1/2}$.

In Table 6.1 the values of the quantity $\sqrt{1 - \frac{\delta}{\delta_T}}$, calculated from equation (6.9) and those from the more precise equation (6.13) are compared.

Correct to 5%, we can apply the equation

$$\alpha_x \approx 0{,}549\lambda\sqrt{\frac{(1 - Pr^{1/3})\, w_0}{ax}}. \tag{6.14}$$

The average value of the heat-transfer coefficient on a plate of length L is equal to

$$\bar{\alpha}=\frac{1}{L}\int_0^L \alpha_x\,dx. \qquad (6.15)$$

Integrating this expression and reducing to dimensionless numbers, we obtain

$$\overline{Nu}\approx 1{,}1\sqrt{(1-Pr^{1/3})\,Pe}\,, \qquad (6.16)$$

where

$$\overline{Nu}=\frac{\bar{\alpha}L}{\lambda} \text{ and } Pe=\frac{w_0 L}{a}.$$

TABLE 6.1

Values of the Quantity $\sqrt{1-\frac{\delta}{\delta_T}}$, as Calculated from Equations (6.9) and (6.13).

$Pr=\frac{\nu}{a}$	0,1	0,05	0,01	0,005	0
$\sqrt{1-Pr^{1/3}}$	0,731	0,795	0,885	0,910	1,0
$\sqrt{1-\frac{2Pr^{1/2}}{Pr^{1/2}+\sqrt{Pr+1{,}56}}}$	0,779	0,835	0,925	0,945	1,0
Discrepancy, %	6	5	5	4	0

Turbulent Boundary Layer Flowing Around A Plate

For a turbulent boundary layer the interpolation expressions (6.3) and (6.4) are unsuitable as temperature and velocity profiles, although the corresponding boundary conditions are still in effect. This is explained by the fact that such profiles do not approximate the logarithmic velocity distribution in a turbulent boundary layer and the corresponding temperature profile with a sufficient degree of accuracy.

The logarithmic velocity profile, which is characteristic for a fully developed turbulent hydrodynamic boundary layer at a plate in definite intervals of the Reynolds number $Re=\frac{w_0 L}{\nu}$ is satisfactorily approximated by a power expression of the form

$$\frac{w_x}{w_0}=\left(\frac{y}{\delta}\right)^n, \qquad (6.17)$$

where the values of n vary with increasing Re ($n=\frac{1}{7},\frac{1}{8},\frac{1}{9}$, etc.).

An approximate distribution of the tangential stresses in the boundary layer at a plate, as obtained by Fediaevskii, is shown in Fig. 6.1. For practical purposes, the turbulent viscosity at the outer boundary of the hydrodynamic boundary layer can be neglected, while in the region near the surface of the plate $\tau=\tau_w$. The latter also explains the applicability of the logarithmic and power velocity distribution laws in the turbulent boundary layer.

When $Pr \ll 1$, thermal conduction outside the hydrodynamic turbulent boundary layer must also be considered. Here the turbulence of the boundary layer can increase the free turbulence of the outer flow in the region $\delta < y < \delta_T$.

In this connection, two analytical schemes are considered below, where we can consider that the true process is contained somewhere in between.

First scheme. It is assumed that the distribution of the tangential stresses in the dynamic layer is linear, i.e.,

$$\tau = \tau_w \left(1 - \frac{y}{\delta}\right) = \frac{c_x}{2} \rho w_0^2 \left(1 - \frac{y}{\delta}\right). \tag{6.18}$$

Taking into account the fact that for a logarithmic velocity profile $\frac{dw}{dy} = \frac{v^*}{\chi y}$, an expression for λ_T can be written as follows:

$$\lambda_T \approx \chi c \gamma \sqrt{\frac{c_x}{2}}\, w_0 y \left(1 - \frac{y}{\delta}\right) \tag{6.19}$$

with $0 < \xi < \beta$ (dynamic layer), and

$$\lambda_T = 0 \tag{6.20}$$

with $\beta < \xi < 1$ (thermal layer).

Here

$$\xi = \frac{y}{\delta_T}, \quad \beta = \frac{\delta}{\delta_T}.$$

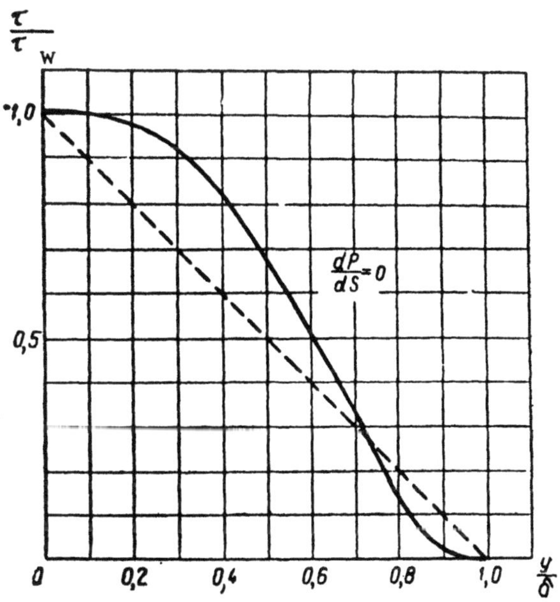

Fig. 6.1. Distribution of the tangential stresses in a boundary layer (with zero pressure gradient).

Let us assume further that the distribution of heat flow within the thermal layer ($0 < \xi < 1$) is linear:

$$q = q_w (1 - \xi). \tag{6.21}$$

If we designate the magnitudes of the heat flow and relative temperature in terms of q_1, q_2 and θ_1, θ_2, respectively, for the regions $\xi < \beta$ and $\xi > \beta$, then we can write the following expressions:

$$\frac{q_1}{q_w} = \frac{\lambda + \lambda_T}{\lambda} \frac{\frac{\partial \theta_1}{\partial \xi}}{\left(\frac{\partial \theta_1}{\partial \xi}\right)_w} = (1 - \xi) \tag{6.22}$$

and

$$\frac{q_2}{q_w} = \frac{\frac{\partial \theta_2}{\partial \xi}}{\left(\frac{\partial \theta_1}{\partial \xi}\right)_w} = (1 - \xi). \tag{6.23}$$

From (6.19) it follows that

$$\frac{\lambda_T}{\lambda} = \frac{\chi c \gamma \sqrt{\frac{c_x}{2}}\, y \left(1 - \frac{y}{\delta}\right)}{\lambda} = B_0 \xi \left(1 - \frac{\xi}{\beta}\right), \tag{6.24}$$

where $B_0 = \frac{c \gamma w_0 \sqrt{\frac{c_x}{2}}\, \delta}{\beta}$.

Now equations (6.22) and (6.23) take on the following form:

$$\theta_1 = \left(\frac{\partial \theta_1}{\partial \xi}\right)_w \int_0^{\xi} \frac{1 - \xi}{1 + B_0 \xi \left(1 - \frac{\xi}{\beta}\right)} d\xi + C_1, \tag{6.25}$$

$$\theta_2 = \left(\frac{\partial \theta_1}{\partial \xi}\right)_W \int_0^{\xi} (1-\xi)\, d\xi + C_2. \qquad (6.26)$$

The integration of these expressions is made with the following boundary conditions, which make it possible to find the unknown constants $\left(\frac{\partial \theta_1}{\partial \xi}\right)_W$, C_1 and C_2:

$$\begin{aligned} \theta_1 &= 0 \quad \text{for} \quad \xi = 0, \\ \theta_2 &= 0 \quad \text{for} \quad \xi = 1, \\ \theta_1 &= \theta_2 \quad \text{for} \quad \xi = \beta. \end{aligned}$$

As a result of integration the following expressions are obtained for the dimensionless temperatures:

$$\left.\begin{aligned} \theta_1 &= \frac{\frac{2-\beta}{\alpha}\ln\left(\frac{\alpha-\beta+2\xi}{\alpha-\beta}\,\frac{\alpha+\beta}{\alpha+\beta-2\xi}\right)+\ln\left(\frac{\alpha-\beta+2\xi}{\alpha-\beta}\,\frac{\alpha+\beta-2\xi}{\alpha+\beta}\right)\frac{\beta}{B_0}}{(1-\beta)^2+\frac{2\beta}{B_0}\,\frac{2-\beta}{\alpha}\ln\frac{\alpha+\beta}{\alpha-\beta}}, \\ \theta_2 &= \frac{2\xi-\xi^2+\beta(\beta-2)+\frac{\beta}{B_0}\,\frac{2(2-\beta)}{\alpha}\ln\frac{\alpha+\beta}{\alpha-\beta}}{(1-\beta)^2+\frac{2\beta}{\alpha}\ln\frac{\alpha+\beta}{\alpha-\beta}}, \end{aligned}\right\} \qquad (6.27)$$

where $\alpha = \sqrt{\beta\left(\beta+\frac{4}{B_0}\right)}$.

It is of interest to note that the magnitude of the derivative of the temperature at the wall $\left(\frac{\partial \theta_1}{\partial \xi}\right)_W$ is constant and equal to 2 for a laminar layer. This quantity in the present case depends on the coordinate x and the Pr number, since in the expression

$$\left(\frac{\partial \theta_1}{\partial \xi}\right)_W = \frac{2}{\frac{\beta}{B_0}\,\frac{2(2-\beta)}{\alpha}\ln\frac{\alpha+\beta}{\alpha-\beta}+(1-\beta)^2} \qquad (6.28)$$

the quantity B_0 is connected with the coordinate x through the functions $c_x = f(x)$ and $\delta = f(x)$, while the quantity β (to a first approximation) is assumed equal to $1.6\,\mathrm{Pr}^{1/2}$. For $\beta \to 0$ (flow of an ideal fluid around a plate) the quantity $\left(\frac{\partial \theta_1}{\partial \xi}\right)_W \to 2$.

It is not difficult to show from (6.28) that in the general case the quantity $\left(\frac{\partial \theta_1}{\partial \xi}\right)_W > 2$.

For an expression which will describe the velocity distribution in the dynamic layer and which is necessary for calculating the integral in equation (6.2), we can assume a power expression

$$\frac{w_x}{w_0} = \left(\frac{y}{\delta}\right)^{1/7}, \qquad (6.29)$$

which, as we know, describes the velocity profile in the turbulent layer with a sufficient degree of accuracy in the interval $Re_x < 10^7$.

Separating the integral in equation (6.2) into two parts, in correspondence with the integration intervals $0 < \xi < \beta$ and $\beta < \xi < 1$, we obtain

$$\int_0^{\delta_T} \frac{w}{w_0}(1-\theta)\, dy = \int_0^{\delta} \left(\frac{y}{\delta}\right)^{1/7}(1-\theta_1)\, dy + \int_{\delta}^{\delta_T} (1-\theta_2)\, dy =$$

$$= \frac{\delta_T}{\beta^{1/7}} \int_0^{\beta} \xi^{1/7}(1-\theta_1)\, d\xi + \delta_T \int_{\beta}^{1} (1-\theta_2)\, d\xi = \delta_T f_0(B_0, \beta). \qquad (6.30)$$

Substituting this expression into the integral equation (6.2) and considering that

$$\frac{\alpha_x}{c\gamma w_0}=\frac{\lambda}{\delta_T c\gamma w_0}\left(\frac{\partial\theta_1}{\partial\xi}\right)_W=\frac{1}{Pe_{\delta_T}}\left(\frac{\partial\theta_1}{\partial\xi}\right)_W, \tag{6.31}$$

we obtain an ordinary differential equation with respect to Pe_{δ_T}:

$$\frac{d}{dx}[\delta_T f_0(B_0,\beta)]=\frac{1}{Pe_{\delta_T}}\left(\frac{\partial\theta_1}{\partial\xi}\right)_W. \tag{6.32}$$

Transforming to the dimensionless coordinate $Pe_x=\frac{w_0 x}{a}$, which is quite convenient in the present case, we can represent equation (6.32) in the following form, assuming the boundary condition $Pe_{\delta_T}/Pe_{x=0}=0$:

$$\frac{d}{dPe_x}=[Pe_{\delta_T} f_0(B_0,\beta)]=\frac{1}{Pe_{\delta_T}}\left(\frac{\delta\theta_1}{\partial\xi}\right)_W. \tag{6.33}$$

The last equation can be solved (for example, by a numerical method), since the form of the functions

$$f_0(B_0,\beta)=\zeta(Pe_x,\beta) \text{ and } \left(\frac{\partial\theta_1}{\partial\xi}\right)_W=\psi(Pe_x,\beta)$$

are known, where in the present case the quantity β is a parameter depending, to a first approximation, only on the Pr number.

The integral which is necessary for determining the function $f_0(B_0,\beta)$, namely $\int_0^\beta \xi^{1/7}(1-\theta_1)\,d\xi$ can be calculated graphically* for every specific value of Pe_x (the interval of practical interest for Pe_x lies within the limits $10^3<Pe_x<10^5$), where the expression for the function $B_0=f(Pe_x)$ can be assumed from equation (6.24).

The magnitude of the local coefficient of resistance of the plate around which the turbulent layer is flowing can be determined from the expression

$$\frac{c_x}{2}=0{,}0289\left(\frac{\nu}{w_0 x}\right)^{1/5}=\frac{0{,}0289}{Re_x^{0,2}}, \tag{6.34}$$

while the thickness of the turbulent layer can be determined from the well-known equation

$$\delta=0{,}37\frac{x}{Re_x^{0,2}}. \tag{6.35}$$

Then

$$B_0=\frac{0{,}0252}{\beta}Pe_x^{0,7}Pr^{0,33}. \tag{6.36}$$

Having obtained the function $Pe_{\delta_T}=f(Pe_x)$ as a result of solving equation (6.33), it is not difficult to go over to a determination of the heat-transfer coefficient α_x:

$$\alpha_x=c\gamma w_0\left(\frac{\partial\theta_1}{\partial\xi}\right)_W\frac{1}{Pe_{\delta_T}}. \tag{6.37}$$

Finally, the Nu number, averaged over the entire plate, can be determined from the expression

---

* The integral $\int_\beta^1(1-\theta_2)\,d\xi$ is easily calculated by analytic means.

TABLE 6.2

Compilation of the Fundamental Analytic Parameters for Calculating the Heat-Transfer Coefficients by the First Scheme

Pe_x	Pe_δ	$\left(\frac{\partial\theta_1}{\partial\xi}\right)_{w\ in}$	$\left(\frac{\partial\theta_1}{\partial\xi}\right)_{w\ I}$	$\left(\frac{\partial\theta_2}{\partial\xi}\right)_{w\ II}$	$B_{o\ in}$	$B_{o\ I}$	$B_{o\ II}$	$f_{o\ in}$	$f_{o\ I}$	$f_{o\ II}$	$Pe_{\delta I}$	$Pe_{\delta II}$	$Pe_{\delta III}$	β_I	β_{II}	β_{III}	Nu_I	Nu_{II}	Nu_{III}
0	0	2,000	2,000	2,000	1,000	1,000	1,000	0,313	0,310	0,310	0	0	0	0	0	0	0	0	0
100	5,85	2,010	2,015	2,015	0,955	0,970	0,970	0,314	0,3099	0,3099	35,75	35,90	35,90	0,1637	0,1630	0,1630	8,10	8,28	8,28
500	20,28	2,045	2,070	2,070	3,060	1,920	1,920	0,3181	0,3095	0,3095	79,70	80,50	80,50	0,2545	0,2520	0,2520	22,28	24,55	24,55
1 000	36,92	2,070	2,135	2,135	4,980	2,420	2,420	0,3212	0,3092	0,3092	112,01	114,7	114,7	0,3290	0,3220	0,3220	33,06	33,48	33,48
2 500	75,45	2,122	2,303	2,303	9,500	3,552	3,552	0,3275	0,3090	0,3090	176,8	183,0	183,8	0,4265	0,4100	0,4100	54,30	55,34	55,34
10 000	233,0	2,240	3,015	2,935	24,900	5,985	6,500	0,3430	0,2925	0,3045	349,95	380,5	380,5	0,6675	0,6130	0,6130	114,9	116,95	123,75
25 000	476,0	2,360	4,085	3,865	47,250	8,590	10,280	0,3580	0,2472	0,2895	343,50	650,0	652,5	0,8780	0,7340	0,7340	193,1	222,6	232,2
50 000	827,0	2,452	5,275	5,040	76,400	11,270	14,950	0,3685	0,1832	0,2638	763,0	1015,0	1013	1,082	0,815	0,815	285,9	364,1	358,2
100 000	1470,0	2,540	6,560	6,810	125,200	14,800	22,950	0,3802	0,1640	0,2300	1084,0	1685	1645	1,353	0,8720	0,8930	422,5	588,3	586,6
250 000	3040,0	2,635	8,550	9,770	238,500	21,595	45,200	0,3915	0,1373	0,2757	1720,0	3615	3085	1,768	0,842	0,985	693,2	1103	1122

Remarks: 1. The index "in" corresponds to the initial analytic values ($\beta = 1.6$, $Pr = 0.5$–0.16). 2. The indices I, II, III, correspond to the first, second, and third approximations.

$$\overline{Nu}=\frac{\bar{\alpha}L}{\lambda}=\frac{L}{\lambda}\frac{\int_0^L \alpha_x\,dx}{L}=\int_0^{Pe_L}\left(\frac{\partial\theta_1}{\partial\xi}\right)_w\frac{1}{Pe_{\delta_T}}\,d\,Pe_x. \tag{6.38}$$

In Table 6.2, the main analytic quantities corresponding to the first, second, and third approximations in calculating the heat transfer by the indicated scheme are presented for the case Pr $\approx$ 0.01 ($\beta \approx$ 0.16). This quantity is almost the same as the real values for Pr for some liquid metals. As evident from the table, the second and third approximations almost coincide.

The results of calculating the heat transfer for Pr = 0.01 (first, second, and third approximations) are shown in Fig. 6.2.

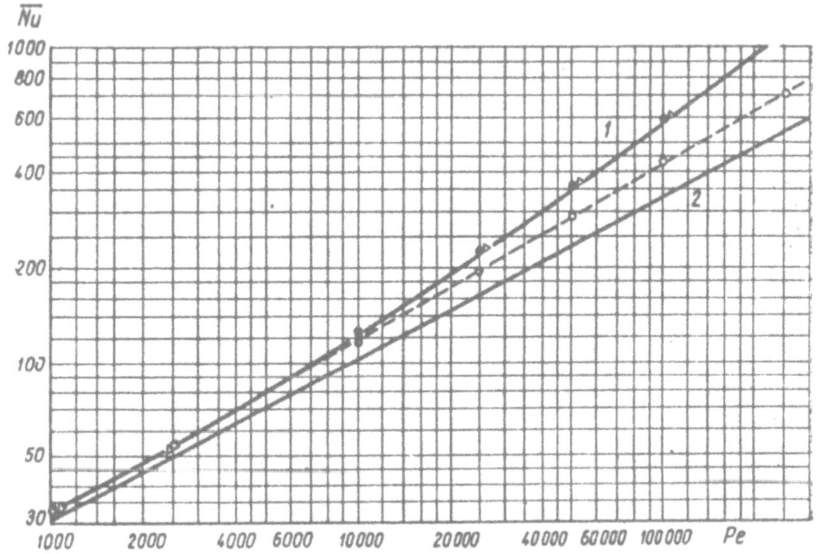

Fig. 6.2. Heat transfer during flow around a plate.

$\tau=\tau_w\left(1-\frac{y}{\delta}\right)$, $Pr=0{,}01$.

O) First approximation; Δ) second approximation; ●) third approximation; 1) turbulent layer $\overline{Nu}$ = 0.38 • Pe$^{0.65}$ (10^3 < Pe < $2\cdot10^5$), 2) laminar layer $\overline{Nu}=1{,}1\sqrt{(1-Pr^{0,33})\,Pe}$.

The function $\overline{Nu}$ = f (Pe), illustrated in this figure (third approximation), can with sufficient accuracy (5 %) be approximated by the equation

$$\overline{Nu}=0{,}38\,Pe^{0{,}65}, \tag{6.39}$$

which is valid in the interval 10^3 < Pe < $2\cdot10^5$.

For comparison, in this same figure the theoretical curve $\overline{Nu}$ = f (Pe) is shown for the case of a laminar layer flowing around the plate, where the applicable equation is (6.16) (Pr = 0.01). As evident from the figure, the increase in the heat-transfer coefficient for the turbulent layer over the laminar case in the interval of Pe numbers of practical interest* amounts to 20-30% on the average.

It should be noted that the distribution in λ_T adopted in the first scheme qualitatively agrees with the analogous distribution for the flow of a fluid with Pr $\ll$ 1 in a tube (in this case the role of the thickness of the dynamic layer is taken by the radius of the tube). In Fig. 6.3, the function $\frac{\lambda_T}{\lambda_{T\,max}}=f\left(\frac{y}{\delta}\right)$ assumed in the

* This interval corresponds to flow around a plate of length approximately 1 m with a flow velocity of the fluid of approximately 1 m/sec.

TABLE 6.3

Compilation of the Fundamental Analytic Parameters for Calculating the Heat-Transfer Coefficient by the Second Scheme

Pe_x	Pe_δ	$\left(\frac{\partial\theta_1}{\partial\xi}\right)_{w\,in}$	$\left(\frac{\partial\theta_1}{\partial\xi}\right)_{w\,I}$	B_0 in	$B_{0\,I}$	f_0 in	$f_{0\,I}$	$Pe_{\delta I}$	$Pe_{\delta II}$	β_I	β_{II}	Nu_I	Nu_{II}
0	0	2,000	2,160	1,000	1,000	0,310	0,3120	0	0	0	0	0	0
100	5,85	2,015	2,260	0,970	0,973	0,3099	0,3067	35,90	37,20	0,163	0,157	8,28	9,49
500	20,28	2,070	2,705	1,920	1,940	0,3095	0,2812	80,50	84,15	0,252	0,241	24,55	26,18
2 500	75,45	2,303	3,600	3,522	3,610	0,3090	0,2353	183,80	199,30	0,410	0,379	55,34	68,62
10 000	233,0	2,935	4,850	6,500	6,500	0,3045	0,1706	380,5	470,0	0,613	0,435	123,8	151,3
50 000	827,0	5,040	7,675	14,950	14,95	0,2638	0,1497	1073	1350	0,815	0,613	358,2	452
100 000	1470	6,810	9,900	22,950	22,30	0,2300	0,1202	1645	2500	0,893	0,588	586,6	682,5

Remarks: 1. The index "in" corresponds to the initial analytic values. 2. The indices I, II, correspond to the first and second approximations.

calculations is compared with a similar function obtained by Isakoff and Drew experimentally in studying the temperature field in the turbulent flow of mercury ($Pr \approx 0.02$).

The absolute values of $\frac{\lambda_T}{\lambda_{T\,max}}$ corresponding to both curves are almost the same.

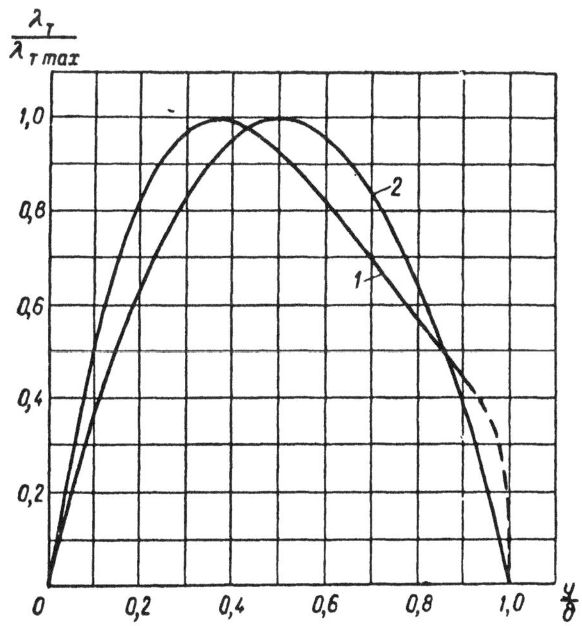

Fig. 6.3. Distribution of the turbulent heat conduction. 1) Analytic function $\frac{\lambda_T}{\lambda_{T\,max}} = 4\frac{y}{\delta}\left(1-\frac{y}{\delta}\right)$; 2) experimental function.

Second scheme. In this scheme it is assumed that the distribution of λ_T within the dynamic layer ($\xi < \beta$) is linear and that λ_T is constant for $\xi > \beta$, i.e., we are essentially considering the condition

$$\tau = \tau_w = \text{const.} \tag{6.40}$$

Actually, if $\tau = \text{const}$, then λ_T cannot revert to zero at the outer boundary of the dynamic layer and the expression for λ_T can be written as follows:

$$\lambda_T = \chi c \gamma \sqrt{\frac{c_x}{2}}\, w_0 y \tag{6.41}$$

for $0 < \xi < \beta$ (dynamic layer), and

$$\lambda_T = \chi c \gamma \sqrt{\frac{c_x}{2}}\, w_0 \delta \tag{6.42}$$

for $\beta < \xi < 1$ (thermal layer).

Further calculations in the present case are similar to those described earlier. Therefore, we shall cite only some numerical data and the final results (Table 6.3). Corresponding data on the heat transfer are shown in Fig. 6.4, where for the sake of comparison the function $\overline{Nu} = f(Pe)$, which was mentioned previously, is plotted for the case $\tau \neq \tau_w$ (third approximation) and for the flow of a laminar layer around a plate. It is evident from the figure that the heat-transfer coefficient for $\tau = \text{const}$ is somewhat higher than for $\tau = \tau_w\left(1 - \frac{y}{\delta}\right)$, which is explained by the large magnitude of the turbulent heat conduction in the first case in comparison with the second. This excess amounts to 20% on the average.

In the interval $Pe = 10^3 - 10^5$ the function $\overline{Nu} = f(Pe)$ is approximated by the equation

$$\overline{Nu} = 0{,}46\, Pe^{0{,}65}. \tag{6.43}$$

Comparing the curves for turbulent and laminar layers cited in Fig. 6.5, we can conclude that turbulization of the boundary layer has a marked effect on heat transfer, in spite of the relatively large role of molecular heat transfer, a fact which is characteristic for media with $Pr \ll 1$.

In Fig. 6.5, the same analytic curves are compared with experimental data, obtained for a flow of liquid with $Pr \approx 0.007$* along the surface of an unbroken section of tubing. The analytic data in general is in agreement with the experimental, in which connection it can be concluded that the initial premises adopted in the calculations are realistic.

* Special calculations, made according to the method described for $Pr = 0.005$ and $Pr = 0.05$, showed that the effect of the Pr number on the heat transfer in the range $0.005 < Pr < 0.05$ (the values of the Pr number for all liquid metals lie within these limits) is extremely insignificant.

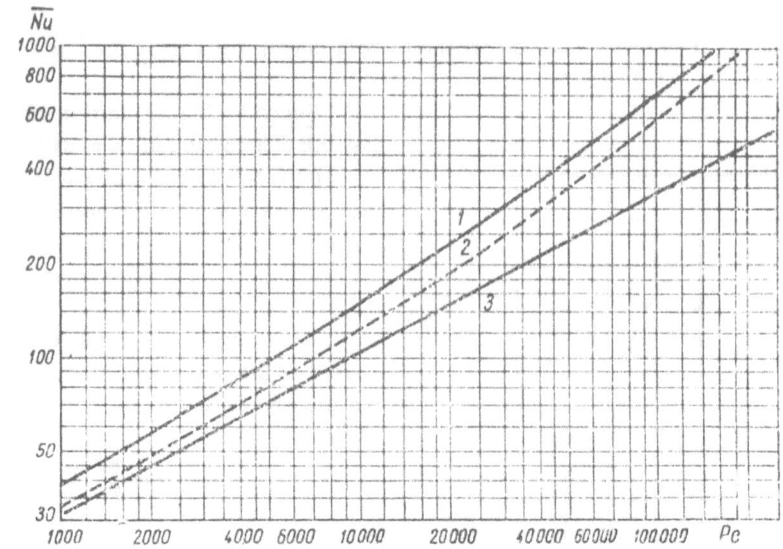

Fig. 6.4. Heat transfer during flow around a plate. $\tau = \tau_w$, Pr = 0.01. 1) Turbulent layer $\overline{Nu} = 0.46 \cdot Pe^{0.65}$ ($10^3 < Pe < 10^5$); 2) $\tau = \tau_w \left(1 - \frac{y}{\delta}\right)$; 3) laminar layer.

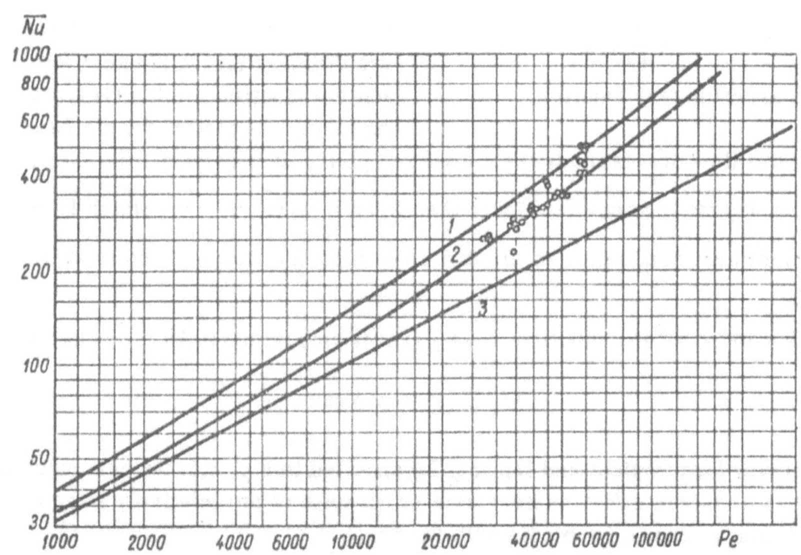

Fig. 6.5. Comparison of the calculations with experimental data. O) Experimental points (Pr ≈ 0.007, longitudinal flow along a single section of tubing); 1) turbulent layer, analytic curve according to the second scheme (Pr = 0.01), 2) turbulent layer, analytic curve according to the first scheme (Pr = 0.01), 3) laminar layer, analytic curve (Pr = 0.01).

Chapter 7

HEAT TRANSFER DURING TRANSVERSE FLOW AROUND CYLINDERS

Heat Transfer During Longitudinal Flow Around A Single Cylinder

In the potential flow in a thin thermal boundary layer the flow velocity along the normal to the surface of an obstacle remains constant and equation (6,1) assumes the form

$$\frac{a}{c\gamma w}=\frac{d}{dx}\left[\int_0^{\delta_T}(1-\theta)\,dy\right]. \tag{7,1}$$

Substituting equation (6,3), the expression for the temperature profile, into the above equation, we obtain

$$\frac{d}{dx}(w\delta)=\frac{20a}{3\delta_T}. \tag{7,2}$$

Here it is assumed that the flow temperature outside the boundary layer and the surface temperature are constant.

Reducing equation (7,2) to the form

$$w\,d\delta_T+\delta_T\,dw=\frac{20a}{3\delta_T}\,dx \tag{7,3}$$

and multiplying both sides of this equation by w, we find that

$$\delta_T^2=\frac{40a}{3w^2}\int_0^x w\,dx. \tag{7,4}$$

Assuming

$$\frac{w}{w_0}=U(\varphi), \tag{7,5}$$

where w_0 is the velocity of the undisturbed flow and φ is an angular coordinate, and bearing equation (6,6) in mind, we obtain

$$Nu=F(\varphi)\sqrt{Pe}, \tag{7,6}$$

where

$$F(\varphi)=0{,}548\frac{w}{w_0}\left(\int_0^{\varphi}\frac{w}{w_0}\,d\varphi\right)^{-1/2}. \tag{7,7}$$

For streamline potential flow around a circular cylinder

$$\frac{w}{w_0}=2\sin\varphi, \tag{7,8}$$

so that, according to equation (7,6),

$$Nu=0{,}99\sqrt{Pe}. \tag{7,9}$$

In these equations $Nu = \frac{\alpha D}{\lambda}$ and $Pe = \frac{w_0 D}{a}$, where D is the diameter of the cylinder.

This solution, obtained by Kruzhilin [56], is very nearly the same as the following exact solution, determined by Piercy and Wihni [78]:

$$\overline{Nu} = 1{,}02\sqrt{Pe}. \tag{7.10}$$

In Table 7.1 analytic results, calculated by Kruzhilin for any part of a cylinder around which a viscous fluid is flowing with Pr $\gtrsim$ 1, are illustrated. For this case

$$Nu = F_1(\varphi)\, Pr^{1/3} Re^{1/2}. \tag{7.11}$$

Comparing the results obtained from these relations with the results obtained in Chapter 6 for longitudinal flow around a plate, we can observe a common behavior in the connection between the Nu and Pe numbers (for Pr $\rightarrow$ 0).

TABLE 7.1
Values of the Functions F (φ) and $F_1(\varphi)$ in Equations (7.6) and (7.11)

$\varphi°$	Potential flow		Viscous flow
	streamline flow	broken flow $1\cdot10^4 < Re < 20\cdot10^4$	$1\cdot10^4 < Re < 20\cdot10^4$
0	1,549	1,484	1,058
10	1,543	1,478	1,058
20	1,526	1,462	1,041
30	1,496	1,435	1,033
40	1,456	1,395	1,018
50	1,404	1,333	0,990
60	1,342	1,252	0,937
70	1,268	1,142	0,874
80	1,187	0,996	0,786
90	1,095	—	0,643
100	0,996	—	—
110	0,868	—	—
120	0,774	—	—
130	0,655	—	—
140	0,530	—	—
150	0,401	—	—
160	0,268	—	—
170	0,135	—	—
180	0,000	—	—

Experimental Data on Average Heat Transfer

The analysis of heat transfer presented above for the case of potential flow around a cylinder makes it possible to perform an order-of-magnitude determination of the Nu number and the nature of its relationship with the Pe number for very small Pr numbers (see Chapter 6).

Without refining further the problem of the effect of the velocity profile on the coefficient of proportionality in equation (7.9) for $\delta_T/\delta \rightarrow \infty$, we shall introduce a correction to the final value of the Pr number in correspondence with equation (6.16):

$$Nu \approx C\sqrt{(1 - Pr^{1/3})\, Pe}. \tag{7.12}$$

The coefficient of proportionality C in the case of streamline potential flow is equal to 0.99; for broken potential flow, assuming a constant level of heat-transfer intensity beyond the break (see below), C = 1.16; for viscous flow, assuming a constant level of heat-transfer intensity beyond the break (if we assume the values of $F_1(\varphi)$ from Table 7.1), $C = 0.80$.

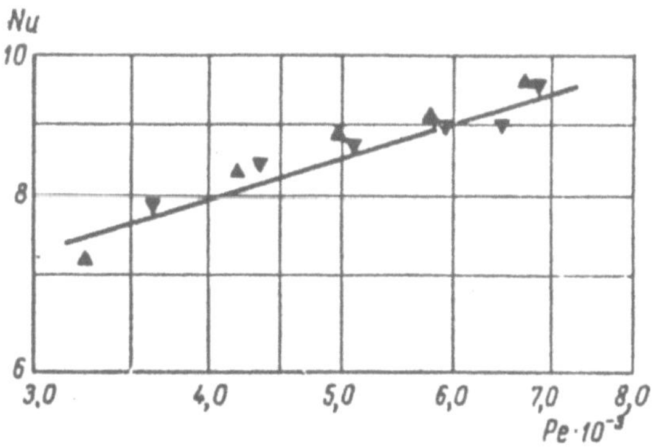

Fig. 7.1. Heat transfer during transverse flow around a single cylinder, assuming Pr ≪ 1. The different points refer to different series of tests.

From the data of Andreevskii, Borishanskii, Kutateladze, and Shneiderman, the number $\overline{Nu}$ for transverse flow around a single cylinder is proportional to the Pe number, raised to the power 0.3-0.5. These values are near the theoretical value, which is equal to 0.5.

In Fig. 7.1, some experimental data on the heat transfer during the transverse flow of a liquid metal with $Pr \approx 0.03$ around a single cylinder are shown. For comparison with the relation (7.12) in terms of the experimental points, a straight line is drawn on the logarithmic scale with a slope of 0.5. This line is defined by the expression

$$\overline{Nu} \approx 0{,}33\, Pe^{1/2}. \qquad (7.13)$$

Substituting the values $C = 0.99$ and $C = 0.5$ into (7.12) for $Pr \approx 0.03$, we obtain values for the proportionality factor associated with $\sqrt{Pe}$ equal to 0.82 and 0.5, respectively. In reference to heavy metals, in the literature [18, 21] it is noted that the time-stabilized values of the heat transfer are 30-40% lower than the original values. A similar effect is also observed in the present case.

Heat-Transfer Intensity Distribution in the Vicinity Surrounding the Cylinder

In Fig. 7.2 curves are presented, showing the local heat transfer in the vicinity of a single cylinder, around which a liquid metal and air are flowing. It is evident from the figure that the high thermal conductivity of the liquid metal and the small Pr number, which reduces the intensity of the turbulent transfer of heat, substantially

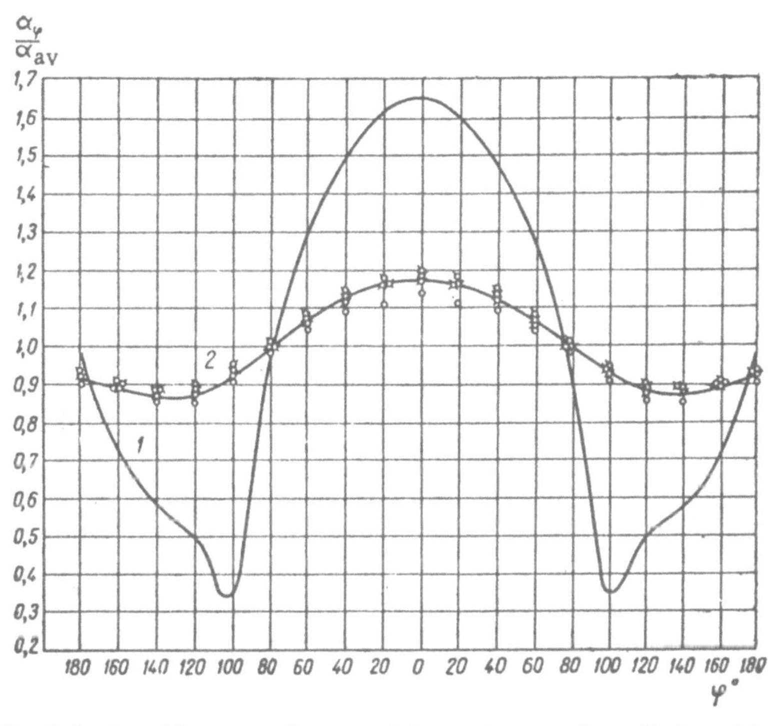

Fig. 7.2. Local heat transfer around the perimeter of a cylinder subjected to transverse flow. 1) air ($Pr \approx 1$), 2) liquid metal ($Pr \ll 1$); O) Pe = 450, ⌀) Pe = 600, ⊠) Pe = 800, ⊠) Pe = 1000.

change the heat-transfer distribution surrounding the cylinder in comparison with the case of the flow of a gas or nonmetallic liquids around a cylinder. The most characteristic factor is the almost constant heat-transfer intensity in the backwash region of a cylinder immersed in a flow of liquid metal. The effect of the Pe number (or Re) on the relative distribution of the coefficient α is very slight. Similar results are obtained also for bundles of cylinders placed in rows and staggered, as shown in Figs. 7.3 and 7.4. A drop in the heat-transfer intensity at the frontal part of the cylinders situated behind the first row of cylinders is observed for the liquid metal flow.

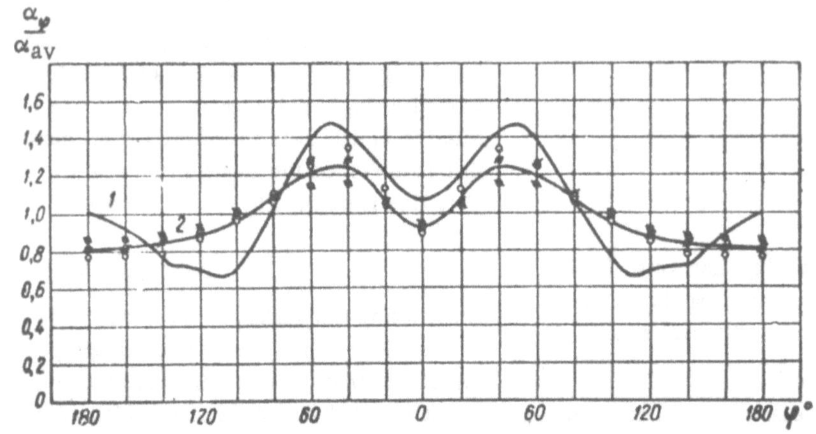

Fig. 7.3. Heat transfer for cylinders placed in a row. 1) Air; 2) liquid metal; ○) Pe = 1200, ∅) Pe = 2000, ⚲) Pe = 3000.

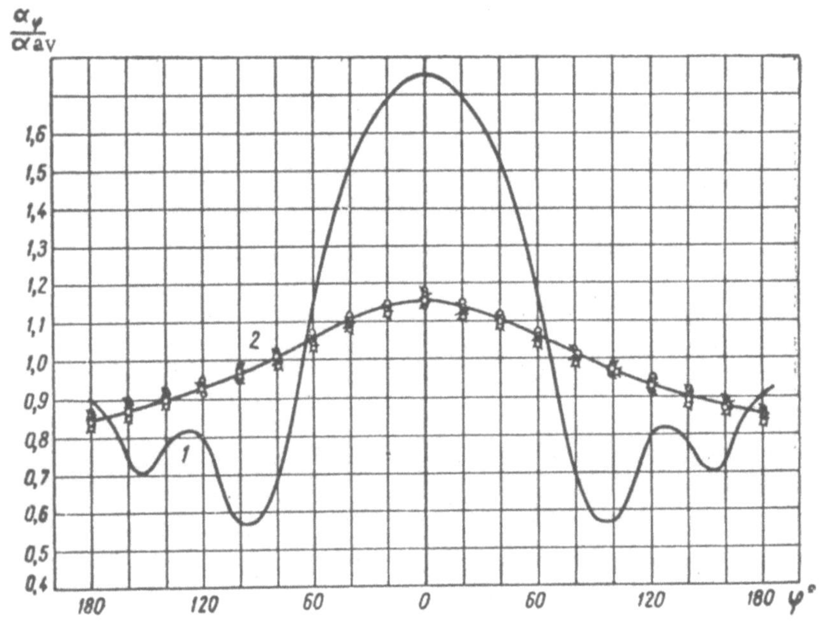

Fig. 7.4. Heat transfer for cylinders placed in a staggered bundle. 1) Air; 2) liquid metal; ○) Pe = 800, ⚲) Pe = 12,000,* ⚲) Pe = 1600, ⌑) Pe = 2000.

* As in original — Publisher's note.

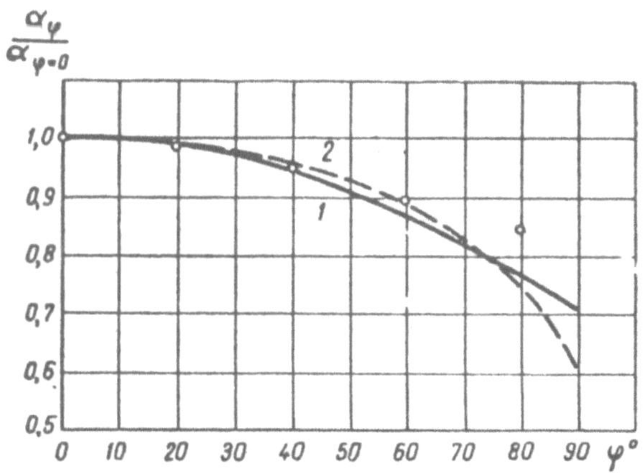

Fig. 7.5. Comparison of experimental and calculated values of α at the frontal part of a cylinder subjected to transverse unbounded flow. 1) Potential flow; 2) viscous flow; o) experimental points.

In Fig. 7.5, the distribution of the quantity α in the vicinity of the frontal part of a single cylinder (obtained experimentally) is compared with the distribution calculated from the data of the first and third columns of Table 7.1. The experimental data are represented by the points associated with the experimental curve of Fig. 7.2. As evident from the figure, the theoretically calculated relative variation in α at the frontal part of a cylinder depends very little on the nature of the flow (potential or viscous) and is almost the same as the experimental data.

Chapter 8

HEAT EXCHANGE DURING FREE CONVECTION

Nature of the Relationship Between the Fundamental Dimensionless Parameters When Pr $\ll$ 1

In the general case the process of heat transfer in free convection is determined by the set of the equations of thermal conductivity, motion, and continuity of the liquid flow. In this case in the equation of motion we must take into account the lift force which arises as a result of the difference in densities of the medium. This force is proportional to the volume coefficient of expansion multiplied by the difference in temperature between some point and a fixed point on the temperature scale. If the process takes place in an unbounded space, then as an initial reference point for the temperature the temperature at a great distance from the heat-exchange surface (temperature of the undisturbed flow) is chosen.

The conditions for uniqueness in the problem to be investigated consist of the shape and dimensions of the disturbing body (hot or cold relative to the initial temperature of the liquid medium), the temperature distribution over the surface of this body and in the liquid, the physical properties of the liquid (λ, a, ν, ρ, β) and the gravitational acceleration g.

The variation in the physical properties of liquid metals as a function of the temperature (except for the density ρ) can be neglected.

Under these conditions Nu is a function of two defining criteria, namely: the Grashof number $Gr = \frac{gl_0^3}{\nu^2}\beta\,\Delta T$ and the Prandtl number $Pr = \frac{\nu}{a}$, where l is the characteristic linear dimension of the body. Thus, the functional dependence of the Nu number can be given in the following general form:

$$Nu = \Phi\left(Pr;\ Gr;\ \frac{l_1}{l_0};\ \frac{l_2}{l_0};\ \ldots\right). \qquad (8.1)$$

In free convection, when essentially the motion of the liquid is completely determined by the process of heat exchange, strictly speaking, it is impossible to consider separately the thermal and hydrodynamic boundary layers. However, general considerations on the relative propagation depth of the effect of molecular friction and molecular thermal conductivity, which is characterized by the quantity ν/a, remain valid even in the present case.

For $\nu/a \ll 1$ the strong effect of the molecular conductivity in free convection is propagated far from the region of the layer near the surface of the body, where there is more or less ordered motion, due to the molecular viscosity. Therefore, retaining from the theory of free convection the usual concept of the determinative effect of molecular heat transport on the process of heat transfer, it should be considered that the velocity field within the greater part of the thermal boundary layer depends mainly on the inertial forces. If, in the equation of motion we use this fact as a basis for dropping terms which involve the effect of molecular viscosity, then we obtain the following set of equations:

$$
\left.\begin{aligned}
&a\nabla^2 t=\frac{\partial t}{\partial \tau}+(\mathbf{w},\ \operatorname{grad} t),\\
&g\beta\Delta t-\frac{1}{\rho}\operatorname{grad} p_d=\frac{\partial \mathbf{w}}{\partial \tau}+(\mathbf{w},\ \operatorname{grad})\ \mathbf{w},\\
&\operatorname{div}\mathbf{w}=0,\\
&-\lambda\left(\frac{\partial t}{\partial n}\right)_W=\alpha\Delta t,
\end{aligned}\right\} \tag{8.2}
$$

where p_d is the dynamic part of the pressure, caused by free convection of the fluid in the thermal boundary layer.

This set of equations, which is characteristic of a certain limiting regime of free convection in a medium with $Pr \ll 1$, gives only one defining criterion:

$$\frac{gl_0^3}{a^2}\beta\Delta T=Gr\,Pr^2. \tag{8.3}$$

From this it follows that for liquid metals the experimental data should be more reducible with the dimensionless function

$$Nu=\Phi\left(GrPr^2;\ \frac{l_1}{l_0};\ \ldots\right), \tag{8.4}$$

rather than the function

$$Nu=\Phi_1\left(GrPr;\ \frac{l_1}{l_0}\right). \tag{8.5}$$

The latter function, which is valid for media having $Pr > 0.5$, is derived on the assumption that in free convection the inertial forces can, as opposed to the molecular friction forces, be neglected.

Theoretical Solution For A Vertical Slab In The Region Of A Fully Developed Laminar Boundary Layer

Free convection will arise around a vertical slab which is immersed in an unbounded flow of a quiescent liquid under the influence of a temperature difference. For sufficiently large slab dimensions (relative to the thickness of the layer of the greatest flow disturbances) the boundary layer which arises can be considered plane. The pressure in each horizontal cross section of such a flow is equal to the hydrostatic pressure in the undisturbed region (since in the boundary layer $dp/dy = 0$).

For a laminar fluid flow the equations of the boundary layer have the form

$$
\left.\begin{aligned}
&a\frac{\partial^2 t}{\partial y^2}=w_x\frac{\partial t}{\partial x}+w_y\frac{\partial t}{\partial y},\\
&g\beta\Delta t+\nu\frac{\partial^2 w_x}{\partial y^2}=w_x\frac{\partial w_x}{\partial x}+w_y\frac{\partial w_x}{\partial y^2},\\
&\frac{\partial w_x}{\partial x}+\frac{\partial w_y}{\partial y}=0,\\
&-\lambda\left(\frac{\partial t}{\partial y}\right)_W=\alpha\Delta t.
\end{aligned}\right\} \tag{8.6}
$$

In these equations the coordinate y is directed along the normal away from the surface of the slab into the fluid, while the coordinate x is directed along the surface of the slab in the direction of the main line of flow of the fluid, where the flow arises as a result of a temperature difference between the surface of the slab and the undisturbed fluid $\Delta t = t_W - t_0$. The origin in the case $t_W > t_0$ is located at the lower end section of the slab, while for $t_W < t_0$ it is located at the upper end.

This problem was first solved theoretically by Lorentz in 1881 for air. Later a method of solution of this set of equations was proposed by Pohlhausen and developed by Saunders [73], Sperrow and Gregg [58], and other researchers.

Equation (8.6) can be transformed by the introduction of the current function $\psi \left(w_x = \frac{\partial \psi}{\partial x},\ w_y = \frac{\partial \psi}{\partial y} \right)$ and the new variables:

a)
$$\eta = \left(\frac{g\beta\Delta t}{4\nu^2} \right)^{1/4} y x^{-1/4},$$
$$f(\eta) = \frac{\psi}{(64\nu^2 g\beta\Delta t x^3)^{1/4}}$$

for t_W = const;

b)
$$\eta = \left(\frac{g\beta q}{5\lambda\nu^2} \right)^{1/5} y x^{-1/5},$$
$$F(\eta) = \frac{\psi\lambda^{1/5}}{(5^4 g\beta q\nu^3 x^4)^{1/5}},$$
$$\theta(\eta) = \left(\frac{g\beta\lambda^4}{5\nu^2 q^4 x} \right)^{1/5} (t - t_0)$$

for q = const.

The mean value of the heat-transfer coefficient is determined from the equation

a)
$$\bar{\alpha} = \frac{1}{L} \int_0^L \alpha_x \, dx$$

for t_W = const;

b)
$$\bar{\alpha} = q \left(\frac{1}{L} \int_0^L (t_W - t_0) \, dx \right)^{-1}$$

for q = const.

The results of the solution to the problem are summarized in Table 8.1.

TABLE 8.1
Value of the Complex $\overline{Nu}Gr^{-1/4}$ For Laminar Flow Around A Vertical Slab
[Exact solution of the set of equations (8.6)]

Pr	$\overline{Nu}\,Gr^{-1/4}$ for $t_W - t_0$ = const	$\overline{Nu}\,Gr^{-1/4}$ for q = const
0,1	0,219	0,237
1	0,535	0,573
10	1,10	1,17
100	2,07	2,18

An approximate solution to the set of equations (8.6) (for the case t_W = const) by means of its transformation into integral equations of the boundary layer and the introduction of approximate velocity and temperature profiles was obtained by Squire [28] in the form of the following equation:

$$\overline{Nu} = 0{,}679 \left(1 + \frac{20}{21 Pr} \right)^{-1/4} (Gr Pr)^{1/4}. \quad (8.7)$$

A similar equation was obtained by Eckert [29]. From this equation, if we assume $Pr = 0.1$, $\overline{Nu}Gr^{-1/4} = 0.212$, which is approximately the same as the exact solution.

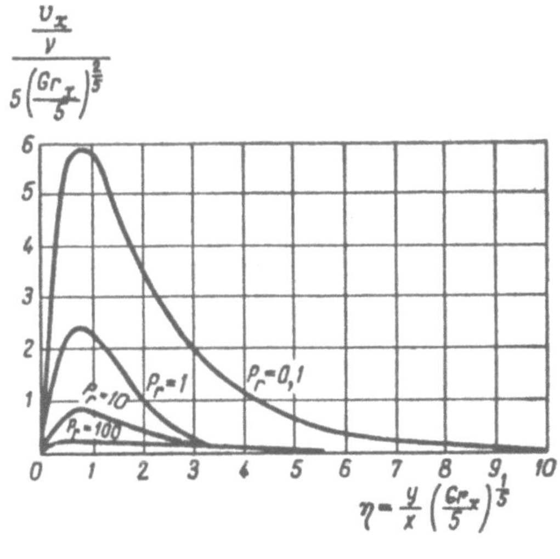

Fig. 8.1. Velocity profile near a vertical plate with a laminar boundary layer.

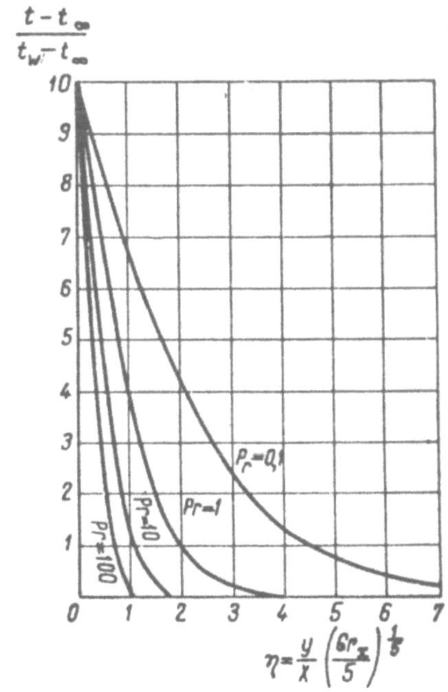

Fig. 8.2. Temperature profile near a vertical plate with a laminar boundary layer.

For $Pr \rightarrow 0$ the ratio $\frac{20}{21Pr} \gg 1$ and equation (8.7) assume the following form:

$$\overline{Nu} = 0{,}679\,(GrPr^2)^{1/4}, \qquad (8.8)$$

which is similar to the general form of the limiting function in (8.3).

In Figs. 8.1 and 8.2 the velocity and temperature profiles around a vertical plate with a laminar boundary layer are illustrated (calculated by Sperrow and Gregg [58]) for the case $q = \text{const}$.

Experimental Data On Free Convection In An Unbounded Volume

In Fedynskii's experiments [74] the heat transfer to mercury, tin, eutectic lead-bismuth, and sodium was studied. The tests were conducted with horizontal cylinders having diameters of 25, 65, and 85 mm and with a vertical slabs having heights of 65 and 101 mm. The results of these tests, along with Saunder's [73] data on mercury, are shown in Fig. 8.3. The curve which corresponds to the Lorentz equation

$$Nu = 0{,}54\,(Gr\,Pr)^{1/4}, \qquad (8.9)$$

in which the coefficient 0.54 is taken from experimental data for media with $Pr \geq 1$ [9, 30, 32, 48] is also plotted.

In Fig. 8.4 the data of Fig. 8.3 are shown as the dependence of the complex $\overline{Nu}Gr^{-1/4}$ on the Pr number. Experimental data obtained by Kirpichev, Voskresenskii, Boiarintsev, Vamsler, Carpenter, Mikheev, Eigenson, Lorentz, Mikheeva, and others for free convection in nonmetallic liquids with $Gr < 10^8$ are also plotted.

These data refer to bodies of various shapes. Consequently, a curve which approximates them gives some averaged effect of the shape on the values of the heat-transfer intensity. On the other hand, from theoretical calculations and a more detailed consideration of the experimental data it follows that the shape of the heat-exchange surface turns out to have some effect on the heat-transfer intensity during free convection. Thus, for example, in the experiments of Germanii it was established that in the region of laminar flow the heat-transfer coefficient for a horizontal tube under almost equal conditions is the same as for a vertical wall of height 2.5 D.

A curve drawn through the experimental points is described by the empirical relation

$$\overline{Nu}Gr^{-1/4} = 0{,}52\,Pr^{0{,}3+0{,}02Pr^{-1/3}}, \qquad (8.10)$$

In [31] the heat transfer during free convection around horizontal cylinders having diameters from 6.3 to 38.1 mm were investigated. The tests were performed with mercury, lead, bismuth, eutectic lead-bismuth, sodium, sodium-potassium alloy, water, toluene, and five different silicates. The authors note that there is no observable effect from the wetting of the heat-exchange surface in the free convection of a nonboiling liquid.

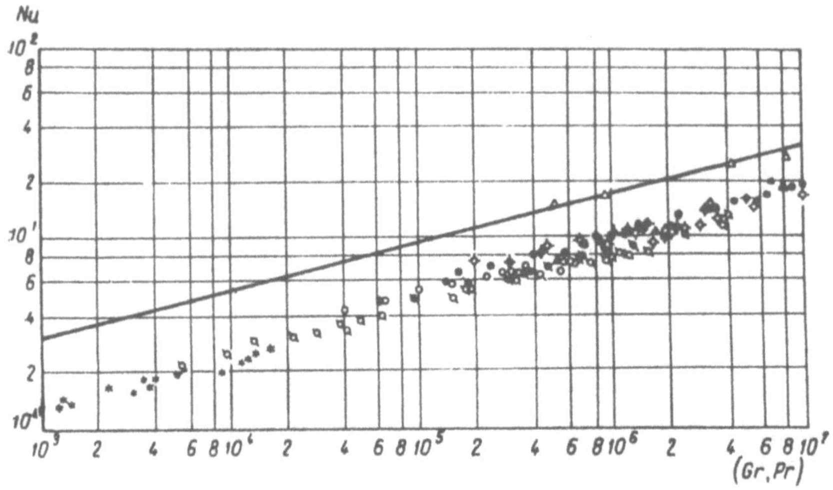

Fig. 8.3. Experimental data on heat transfer for metallic liquids.
△○◊ -◇- ✳) Cylinders; ▲◆● -◆-) slabs; △) water; ▲○●) mercury; ◊ ◆) tin; -◇- -◆-) eutectic lead-bismuth; ✳) sodium.

In processing the experimental data the results of Adams, Ackerman, et al., [32] were also used.

The reduced experimental data are cited in Fig. 8.5. For the region of laminar flow the authors recommend the equation

$$\overline{Nu}Gr^{-1/4} = 0{,}53\left(\frac{Pr^2}{0{,}952 + Pr}\right)^{1/4} \tag{8.11}$$

The heat-transfer law for free convection varies for very large values of the Gr number independently from the value of the Pr number. Physically, this variation is connected with the fact that the laminar behavior of the flow around a heating surface as a whole is broken up and so-called thermal turbulence arises. For this regime of flow around the surface there exists a viscous layer from whose outer side turbulent eddies arise, disrupting the flow. The nature of the motion of the liquid on the average (statistically) becomes identical for different portions of the heat-exchange surface and the heat-transfer coefficient ceases to depend on the dimensions of the body. This is described by the criterion equation

$$Nu = \varphi(Pr)\, Gr^{1/3}, \tag{8.12}$$

It is customary to consider that for media with Pr > 0.5 the transition from equation (8.11) to equation (8.12), which in this case has the form

$$Nu = 0{,}13\,(GrPr)^{1/3}, \tag{8.13}$$

is effected for $GrPr > 2 \cdot 10^7$.

There is no basis for assuming that for very small Pr numbers the combined effect of both defining criteria on the transition from one regime of free convection to the other does not change.

According to Fedynskii's data [74], this transition is effected for values of $Gr > 1.5 \cdot 10^8$ independently from the value of the Pr number.

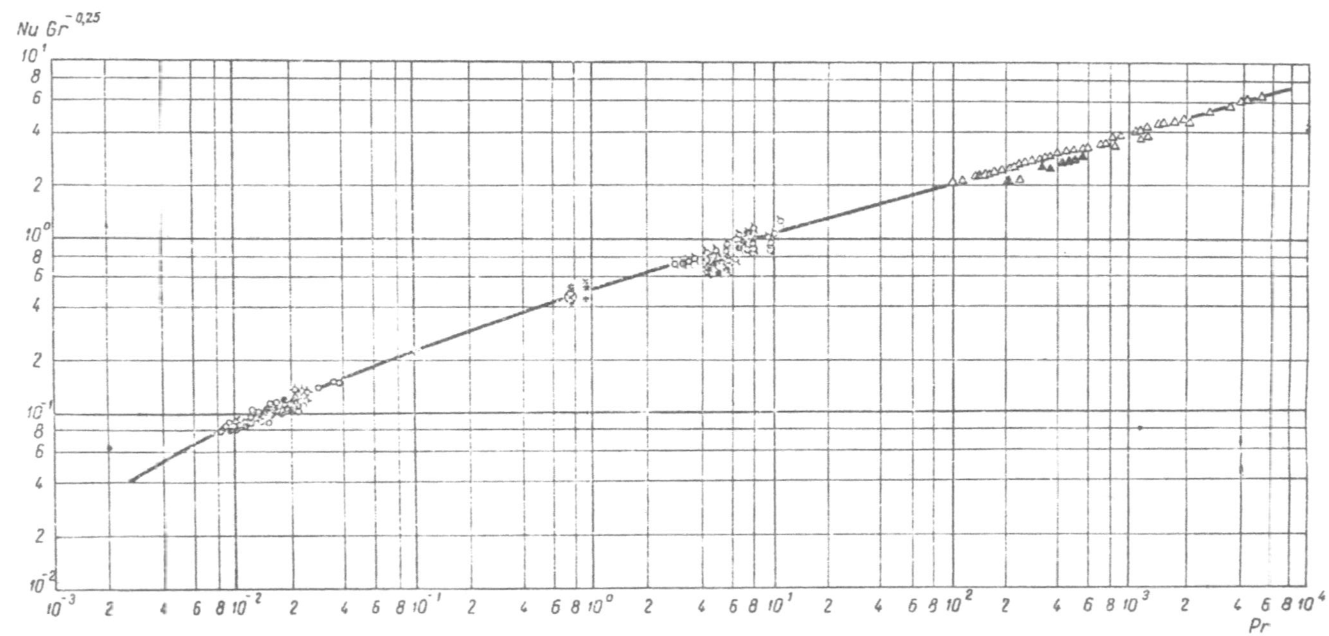

Fig. 8.4. Experimental data on heat transfer for various liquids. ○ ⍉ × △ ⌖) Tubes; ●▲◖) slabs; ✳○) spheres; ●◖◗) water; ✳×) gases (air); △▲) mineral oils; ○●) molten metals (mercury, tin, lead-bismuth alloy, sodium); ⌖) mercury [73].

In the experiments of Hyman, Bonilla, and Ehrlich [31] the onset of turbulence depended not only on the product GrPr, but also on the absolute value of the diameter of the cylinder. The latter testifies to the fact that the authors either did not observe the correct dimensionless function for determining the transition effect, or in their experiments acquired errors in the observations during this phenomenon.

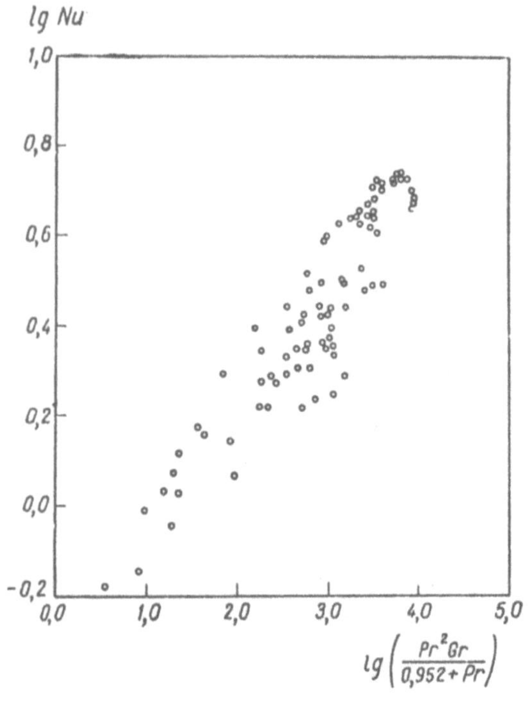

Fig. 8.5. Generalized experimental data on heat transfer for mercury, eutectic lead-bismuth, lead, bismuth, sodium, and sodium-potassium alloy (44% K).

Fedynskii made an analysis of the data and showed that the function $\varphi(\text{Pr})$ in equation (8.12) almost coincides with the right-hand side of the empirical relation (8.10).

On this basis it is possible to assume a unit function for the Pr number in both regions of flow during free convection. Bearing in mind the fact that a calculation of the heat transfer in free convection is not particularly exact, owing to the action of random external perturbations, it is convenient to assume a simpler expression for $\varphi(\text{Pr})$. Proceeding from the theoretical and experimental data presented above, the following function is recommended:

$$Nu \approx \Phi\left(\frac{Pr^2Gr}{1+Pr}\right). \qquad (8.14)$$

This function can be divided into separate intervals, which in turn can be approximated by power equations of the type

$$Nu \approx C\left(\frac{Pr^2Gr}{1+Pr}\right)^n. \qquad (8.15)$$

For $10^2 < \text{Gr} < 10^8$, $C = 0.67$ and $n = 1/4$, for $\text{Gr} > 10^8$, $C = 0.16$ and $n = 1/3$.

The values of the coefficients C are chosen in such a way that the most authentic values of the Nu number, both for $\text{Pr} \ll 1$, and for Pr of the order of unity, are obtained.

Free Convection In A Bounded Space

Fully developed convection in a bounded space (closed cylinder, tubular system, duct, etc.) depends on the shape of the body and on the distribution of the temperature over its separate elements. Thus, for example, in free convection between two plane-parallel long plates on both plates the identical temperature can be maintained or one of the plates can be heated more than the other. Fully developed free convection will obviously not be the same in both cases.

There is a shortage of data as yet on this type of flow, and what there are cannot always be mutually compared. Therefore, we shall not consider this case further. However, in the Literature Cited papers in which in one form or another this problem is considered are given. The general indication reduces to the fact that the principle of generalizing the data on heat transfer in free convection, as presented in detail above and expressed in the form of the relation (8.14), can also be applied in the present case.

Chapter 9

HEAT TRANSFER DURING VAPOR CONDENSATION

Thermal Resistance of a Condensate Film When Pr ≪ 1

Upon cooling the vapors of liquid metals, their condensate, as for the vapors of other substances, is precipitated on the cooling surface in drop form (drop condensation) or in the form of a thin solid film (film condensation). Drop condensation is observed in the case of poor wetting of the cooling surface by the condensate and, as opposed to the majority of nonmetallic liquids, is most characteristic of the type of condensate produced by metallic heat-transfer media.

The high thermal conductivity of the vapor condensate of a metal also leads to an effective redistribution of the thermal resistances inherent in the condensate and the diffusion region around the surface of the condensate. In this connection, the assumption ordinarily made in film condensation theory that the temperature at the surface is practically equal to the saturation temperature in the flow nucleus of the vapor is not justified in the case of the condensation of metallic vapors.

In the laminar flow of a condensate film, regardless of the nature of the substance, Nusselt's theory is quite satisfactory. According to this theory, the thermal resistance of the film of a condensate which is flowing over a vertical wall is equal to

$$R_x^* = \frac{t_f - t_W}{q} = \frac{\delta}{\lambda} = \left(\frac{4\mu q x}{\lambda^3 \gamma^2 r}\right)^{1/4} = 1{,}59\left(\frac{\nu^2}{g\lambda^3}\right)^{1/3} Re_f^{1/3}, \tag{9.1}$$

where R_x^* is the thermal resistance of the condensate film at a cross section which is x meters distant from the top of the wall, in $m^2 \cdot hour \cdot {}^\circ C/kcal$, r is the latent heat of vaporization in kcal/kg, λ is the thermal conductivity of the condensate in $kcal/m \cdot hour \cdot {}^\circ C$, γ is the volumetric weight of the condensate in kg/m^3, μ is the dynamic viscosity coefficient of the condensate in $kg \cdot hour/m^2$, q is the thermal flow density in $kcal/m^2 \cdot hour$, t_f is the temperature of the condensate at its boundary with the vapor, in °C, t_W is the temperature of the condensate at its boundary with the cooling surface, in °C, and δ is the thickness of the condensate film in m.

The laminar regime of flow of the film is observed when

$$Re_f = \frac{G}{g\mu} < 100, \tag{9.2}$$

where $G = \frac{qx}{r}$ is the amount of the condensate deposited on the cooling surface over a width of 1 m, in $kg/m \cdot hour$, g is the gravitational acceleration m/hr^2.

In the general case of turbulent flow of the condensate film we can write:

$$q = -\lambda\left(1 + \varepsilon Pr \frac{\mu_T}{\mu}\right)\frac{dt}{dy}. \tag{9.3}$$

Here, if we neglect supercooling of the condensate, the heat flow q can be considered constant in the direction of the coordinate y.

The distribution of the tangential stresses in the film which is running down a vertical wall under the influence of gravity is determined by the equation

$$\tau = (\gamma - \gamma'')(\delta - y) = \tau_W \left(1 - \frac{y}{\delta}\right), \tag{9.4}$$

where γ'' is the weight by volume of the vapor.

Hence, for the region of logarithmic velocity distribution we have

$$\frac{dw}{dy} = \frac{v^*}{\chi y},$$

$$\frac{\mu_T}{\mu} = \frac{\tau}{\mu \frac{dw}{dy}} = \frac{\chi y}{\nu}\left(1 - \frac{y}{\delta}\right)\sqrt{g\left(1 - \frac{\gamma''}{\gamma}\right)\delta}. \tag{9.5}$$

As explained in the preceding chapters, near the viscous sublayer the quantity λ_T in liquid metals is very small, which makes it possible to assume a single law for determining this quantity over the entire depth of the flow. Bearing this fact in mind, let us substitute the value μ_T/μ from equation (9.5) into equation (9.3) and integrate the equation so obtained over the interval from $y = 0$ to $y = \delta$. We then obtain

$$R_x^* = \frac{t_f - t_W}{q} = \frac{\delta}{\lambda}\int_0^1 \frac{d\xi}{1 + \varepsilon Pr \chi \left(\frac{g\delta^3}{\nu^2}\right)^{1/2}(\xi - \xi^2)}, \tag{9.6}$$

where $\xi = \frac{y}{\delta}$. Here we recall that ordinarily in practice $\gamma'' \ll \gamma$.

The thickness of the condensate film in the turbulent flow regime is defined as a function of the number Re_f by the same formula as in the case of the condensate of nonmetallic vapors. This relation is shown graphically in Fig. 9.1.

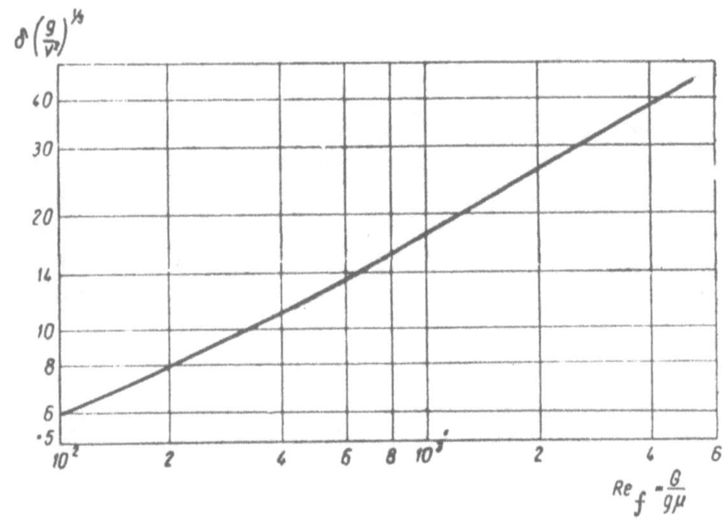

Fig. 9.1. Variation in the thickness of the condensate film.

After performing the integration in equation (9.6) and making some simple transformations, we can determine the local thermal resistance of the condensate film in the region of turbulent flow [40]:

$$R_x^*\lambda\left(\frac{g}{\nu^2}\right)^{1/3}=2\left(\frac{\nu^2}{g\delta^3}\right)^{1/6}(\chi_{\bar{\varepsilon}}Pr)^{-1}\left[\frac{\chi_{\bar{\varepsilon}}Pr\left(\frac{g\delta^3}{\nu^2}\right)^{1/2}}{4+\chi_{\bar{\varepsilon}}Pr\left(\frac{g\delta^3}{\nu^2}\right)^{1/2}}\right]^{1/2}\times$$

$$\times\ln\frac{1+\left[\frac{4+\chi_{\bar{\varepsilon}}Pr\left(\frac{g\delta^3}{\nu^2}\right)^{1/2}}{\chi_{\bar{\varepsilon}}Pr\left(\frac{g\delta^3}{\nu^2}\right)^{1/2}}\right]^{1/2}}{1-\left[\frac{4+\chi_{\bar{\varepsilon}}Pr\left(\frac{g\delta^3}{\nu^2}\right)^{1/2}}{\chi_{\bar{\varepsilon}}Pr\left(\frac{q\delta^3}{\nu^2}\right)^{1/2}}\right]^{1/2}}, \qquad (9.7)$$

where $\bar{\epsilon}$ is some "effective" mean value of the coefficient ϵ.

In Fig. 9.2 are shown results of calculations of thermal resistances from equations (9.1) and (9.7) for certain values of the Pr number which are characteristic for liquid metals. These graphs clearly show that for $Pr \rightarrow 0$ turbulence leads to an increase in the thermal resistance of the condensate film, rather than to its reduction, as is the situation with nonmetallic liquids.

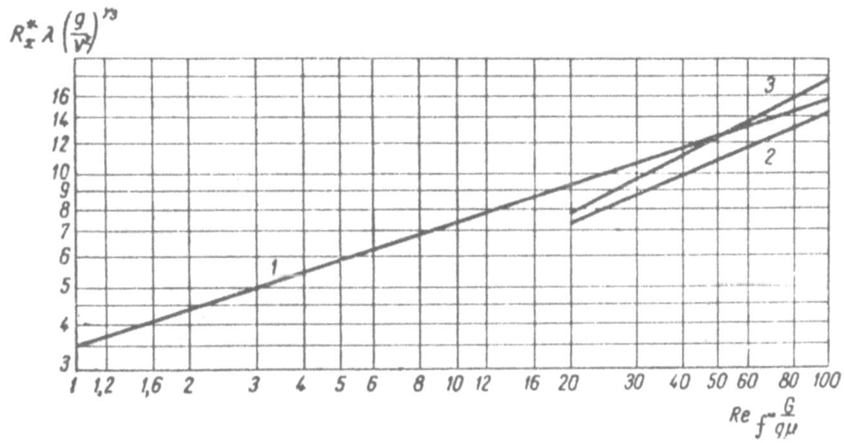

Fig. 9.2. Variation in thermal resistance for various Re numbers.
1) Laminar flow of a condensate film; 2) turbulent flow, with Pr = 0.01;
3) turbulent flow, with Pr = 0.005.

This phenomenon is explained by the fact that when the flow becomes turbulent, the frictional drag in the condensate film is increased and its thickness increases markedly in comparison with the amount which would normally take place for the same value of the Re number, were the laminar regime in effect. This increase in the friction for media with Pr > 1 leads to an even greater increase in the turbulent heat conduction, as a result of which the thermal resistance of the condensate film of nonmetallic liquids is considerably diminished when it becomes turbulent.

In the condensation of the vapors of metals the opposite pattern is observed. In this case, turbulization also leads to an increase in the film thickness, as in the case of nonmetallic liquids. However, due to the fact that Pr < 1, the relative increase in the heat conduction turns out to be smaller than the increase in the thickness of the film. In the limiting case, when Pr = 0, the effective thermal conductivity of the film generally does not change, and the increase in its thickness due to turbulent drag leads to a decrease in the thermal resistance.

Some Considerations of the Mechanism of Drop Condensation

Drop condensation arises on a cooling surface which is not wetted by the condensate and is characterized by the formation on it of a large number of drops, which, upon attaining a definite size, roll down.

Figures 9.3 and 9.4 illustrate photographs of the drop condensation of water on a brass tube covered with a thin layer of grease (from Kutateladze's data [40]) and the drop condensation of mercury on a vertical carbon steel tube (from Gel'man's data [81]). From the photographs it is evident that the pattern of the process in both cases is effectively the same. At points of the cooling surface which are more susceptible to the process of condensation (the so-called "condensation centers") there arise drops of the condensate which occupy a comparatively small portion of this surface. In the intervals between the condensation centers the vapor comes in direct contact with the cooling surface and is possibly separated from it by a condensation layer whose thickness is commensurable with the dimensions of molecular layers.

Fig. 9.3. Drop condensation of water on a brass tube coated with a thin film of grease.

The diameter of the drops breaking away from a vertical cooling surface in a slowly moving vapor can be determined from the equation

$$D_0 = 0{,}0180\,\theta\sqrt{\frac{\sigma}{\gamma-\gamma''}}, \qquad (9.8)$$

where θ is the limiting wetting angle in degrees and σ is the coefficient of surface tension of the condensate in kg/m. In Gel'man's experiments on the condensation of mercury vapor a certain increase in the breakaway diameter of the drops with an increase in the heat flow was observed, along with an almost constant rate of drop formation at the corresponding condensation center (Fig. 9.5).

At the condensation surface a vapor pressure corresponding to the temperature of this surface, i.e., one which is less than in the main portion of the vapor space, is established. Under the influence of this difference in pressures there arises a vapor flow, which is directed along the normal to the condensation surface. If we designate the mass flow rate of this current by g_c (in kg/ m^2 • hr), then the corresponding heat flow will be equal to

$$q = rg_c \ \text{kcal/m}^2\cdot\text{hour} \qquad (9.9)$$

Fig. 9.4. Drop condensation of mercury on a vertical carbon steel tube.

In a pure vapor the mass flow rate will be connected with the pressure drop by the familiar relation

$$g_c = \sqrt{2g\gamma''\Delta p}. \tag{9.10}$$

On the other hand, the pressure drop Δp in a saturated vapor is connected with the temperature difference Δt by the familiar thermodynamic expression*

$$\Delta p = \frac{427\, r\gamma\gamma''\Delta t}{T''(\gamma-\gamma'')}, \tag{9.11}$$

where $T'' = 273 + t''$ is the absolute temperature of saturation.

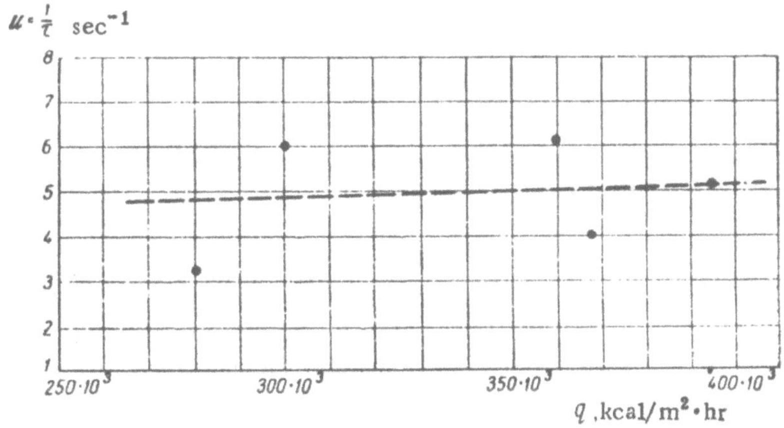

Fig. 9.5. Drop formation rate in the condensation of mercury vapor as a function of the heat load.

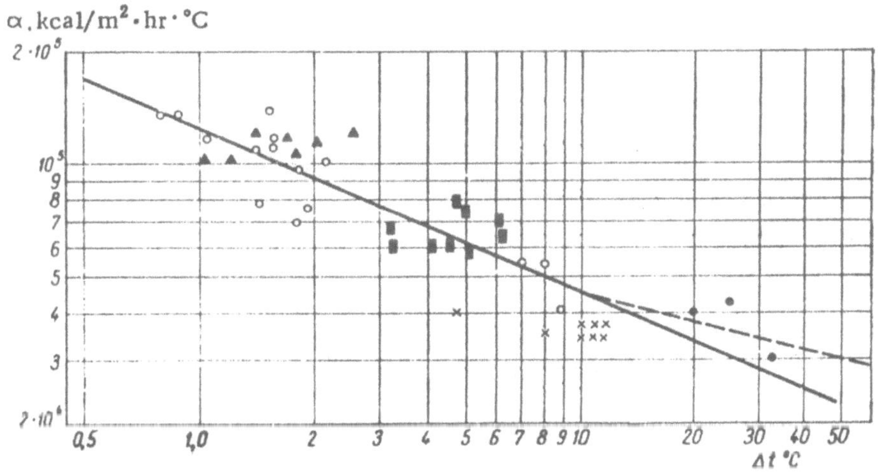

Fig. 9.6. Dependence of the heat-transfer coefficient α on Δt during the drop condensation of water vapor, according to the data of various authors (p ≈ 1 atmos.). ×) Gudemchuk; ○) Kutateladze; ●) Schmidt; ▮) Naigle; ▲) Gnam.

---

* Assuming that $\Delta p = \frac{dp}{dt''}\Delta t$.

Substituting the value of Δp from (9.11) into (9.10), we find that in this case

$$\alpha = \frac{q}{\Delta t} = (r\gamma'')^{3/2} \sqrt{\frac{854\, g\gamma}{T''(\gamma - \gamma'')\Delta t}}, \tag{9.12}$$

i.e., the heat-transfer coefficient for drop condensation should substantially diminish with an increase in the temperature drop between the vapor and the wall.

As evident from Fig. 9.6, this derivation is supported by the experimental data of various authors on the drop condensation of water. However, the absolute values are much smaller than those which follow from equation (9.12). This is explained by the fact that in practical situations the vapor always contains a few, though perhaps insignificant admixtures of foreign gases, which do not condense in the specified temperature interval. These gases accumulate near the condensation surface, so that the pressure of the vapor-gas mixture remains the same everywhere. The vapor-gas boundary layer formed reduces the condensation rate, which in this case is determined by the diffusion intensity of the vapor.

Heat Transfer During the Condensation of Mercury Vapor

The heat transfer of mercury vapor during its condensation on steel tubes has been investigated in considerable detail recently [3]. The condensation and heat-transfer intensity are almost independent from the orientation of the tube relative to the horizontal plane.

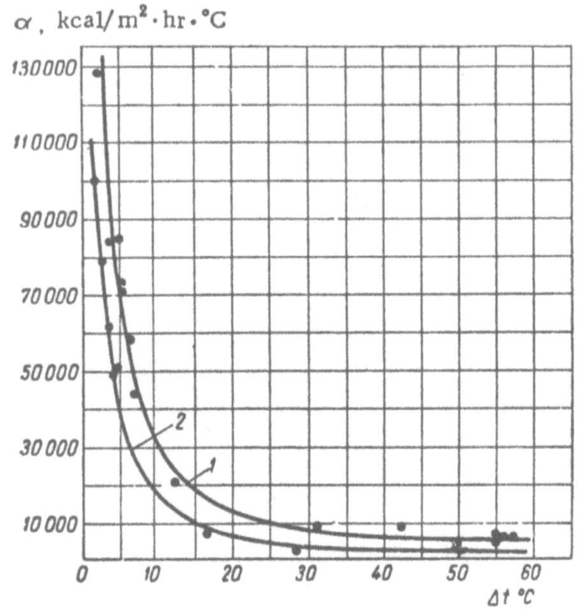

Fig. 9.7. Dependence of the heat-transfer coefficient on the temperature drop during the condensation of pure mercury vapor on a vertical tube. 1) Pressure of 0.75–0.85 atmos.; 2) pressure of 0.13–0.17 atmos.

Figure 9.7 shows the dependence of the heat-transfer coefficient in the condensation of mercury vapor for two pressures in the condenser and an approximately constant velocity of the vapor. The heat-transfer coefficient drops off so sharply with increasing temperature drop that the heat flow q for a specified pressure and velocity of the vapor remains almost constant.

An increase in pressure increases the heat-transfer intensity, which in form corresponds to equation (9.12). However, the quantitative effect of the pressure in the tests is considerably poorer.

The results of Gel'man's experiments [81] with almost pure mercury vapor ($K < 2\%$) are expressed by the empirical equation

$$q = 1{,}2 \cdot 10^5 \sqrt[3]{p}\,(1 + \sqrt[3]{\gamma'' w''}) \text{ kcal/m}^2\cdot\text{hr} \tag{9.13}$$

where p is the pressure of the mercury vapor in the condenser, in atmos., and w'' is the flow rate of the vapor in m/sec.

In Fig. 9.8 the experimental data is compared with calculations made according to equation (9.13).

According to these tests, in the gravitational condensation of air in the mercury vapor from $K = 2$ to 12%,

$$q = 1 \cdot 10^5 \sqrt[3]{p}\,(1 + \sqrt{\gamma_{mix}\, \omega_{mix}})\, K^{-0.2}. \tag{9.14}$$

The indicated tests were conducted over a pressure interval from 0.1 to 1.2 atmos., with temperature drops from 1 to 90° C, and mass flow rates of the vapor or vapor-air mixture from 3 to 45 kg/m²·sec.

Gol'man's tests show that the heat flow density is practically independent from the temperature drop during the drop condensation of mercury vapor, i.e., according to these data, $\alpha \approx \Delta t^{-1}$. Thus, even in the drop condensation of metallic vapors the heat-transfer coefficient is substantially reduced with an increase in the temperature drop, which is in qualitative agreement with the theoretical derivation. However, the complete independence of the heat flow from Δt is not sufficiently clear. Evidently, the power exponent of Δt is actually slightly less than unity.

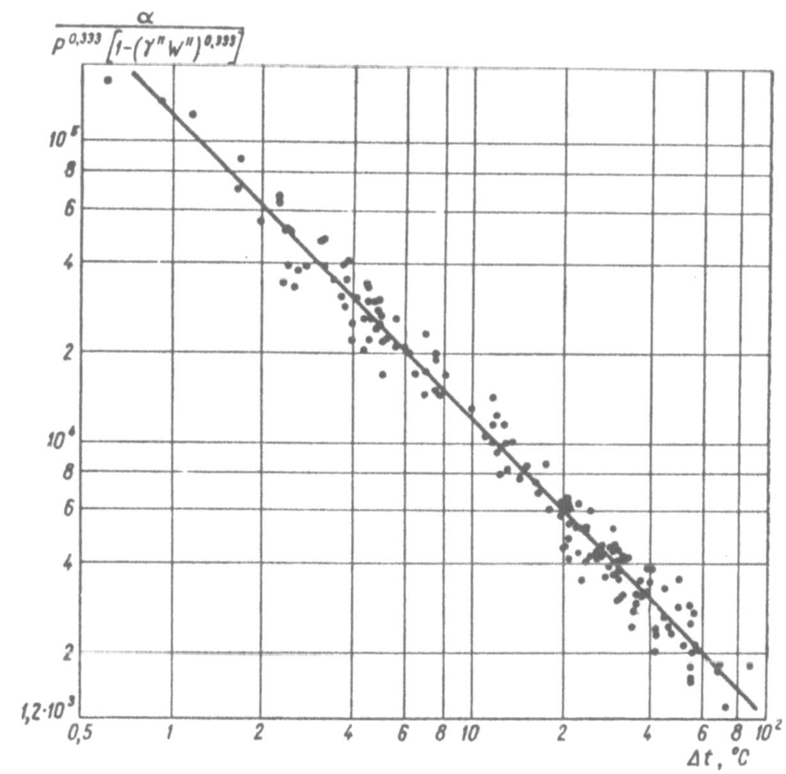

Fig. 9.8. Results of experiments on the condensation of pure mercury vapor.

Heat Transfer During the Condensation of Sodium Vapor

At the present time there exists no comprehensive data on the heat transfer during the condensation of the vapors of metals other than mercury. The heat-transfer coefficient in the condensation of sodium vapors is given in order of magnitude in Table 9.1.

TABLE 9.1
Values of the Heat-Transfer oefficient in the Condensation of Sodium Vapor
(From the data of Bonilla and Meyer)

Saturation temperature t", °C	631	725	866
Heat flow q, kcal/m^2•hour	$1.5 \cdot 10^5$	$1.92 \cdot 10^5$	$2.66 \cdot 10^5$
Average heat-transfer coefficient $\bar{\alpha}$, kcal/m^2•hr•°C	$5.55 \cdot 10^4$	$5.85 \cdot 10^4$	$6.50 \cdot 10^4$
Average heat-transfer coefficient [from equation (9.1)] $\bar{\alpha}$ kcal/m^2•hr•°C	$84.5 \cdot 10^4$	$74.0 \cdot 10^4$	$61.2 \cdot 10^4$

These data refer to a tube of diameter 12.5 mm, length 15 mm, and inclined at an angle of 45°. In the calculation from equation (9.1) the usual corrections for the angle of incline are introduced and the result is averaged over the length of the condensation surface.

The experimental heat-transfer coefficients turn out to be considerably smaller than those calculated from the thermal resistance of a condensate film. A similar result is obtained in the condensation of mercury vapor. From this it can be stated that in the condensation of metallic vapors the main thermal resistance is concentrated at the boundary between the vapor and the condenser, while the thermal resistance of the condensate is relatively small. Consequently, the type of condensation (film or drop) has little effect on the heat-transfer intensity in the condensation of metallic vapors.

Chapter 10

HEAT EXCHANGE DURING BOILING

Fundamental Considerations of the Mechanism of Boiling

In Fig. 10.1, the dependence of the heat-transfer coefficient on the temperature drop during the boiling of water and a magnesium amalgam of mercury on a horizontal tube under the conditions of free convection (i.e., with boiling on a heating surface immersed in a large volume of liquid), is shown in logarithmic coordinates. For both liquids (metallic and nonmetallic) the behavior of this dependence is essentially the same.

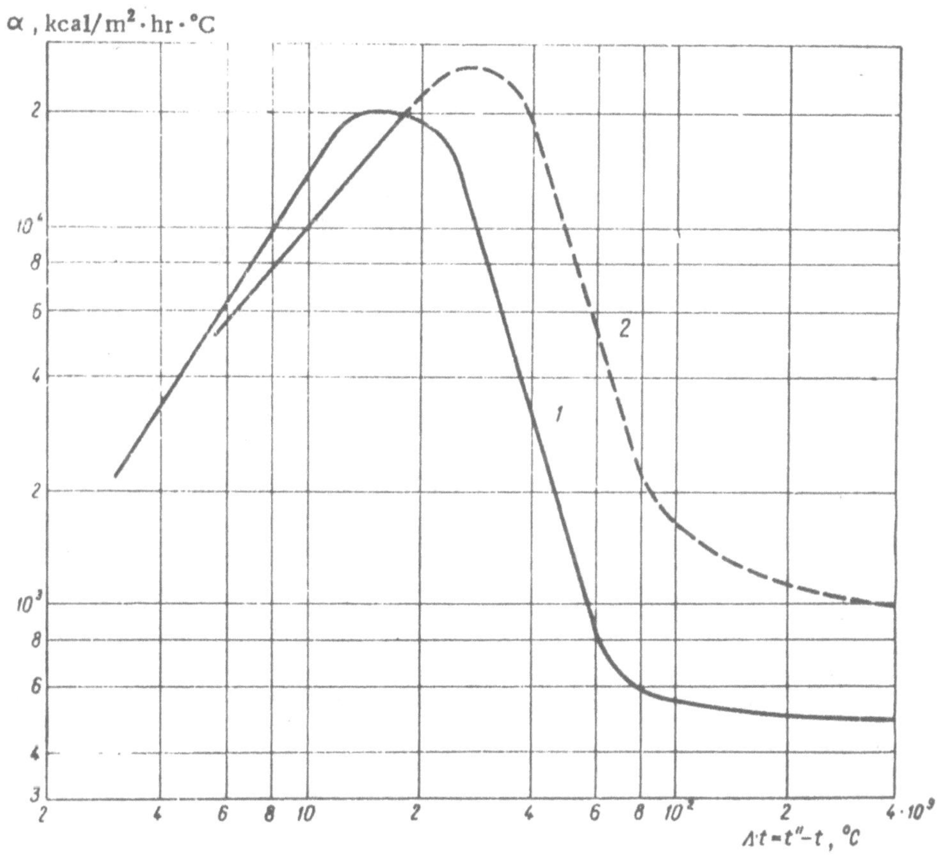

Fig. 10.1. Heat-transfer during the boiling of water and a magnesium-mercury amalgam. 1) Water; 2) amalgam.

Three pronounced regions of temperature change are evident, where the heat-transfer law becomes considerably different.

The left-hand, sharply rising branch of the curve is represented by the heat-transfer law for boiling in the bubbling regime.

The right-hand, nearly horizontal branch of the curve is represented by the heat-transfer law for boiling in the pellicular regime.

The intermediate, falling branch of the curve is represented by the heat-transfer law for boiling in the transition regime.

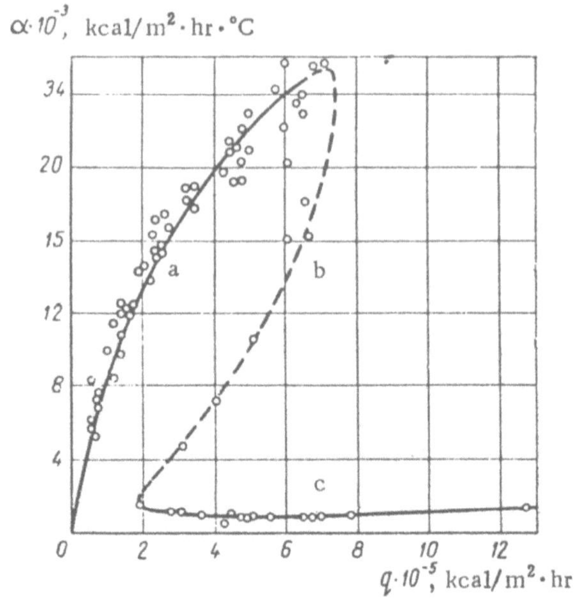

Fig. 10.2. The function $\alpha = f(q)$ during the boiling of water. a) Bubbling regime; b) transition zone; c) pellicular regime.

With a smooth increase in the temperature difference the heat-transfer coefficient, after attaining a maximum, more or less falls off smoothly. In this case, after the maximum value of α the heat flow density also diminishes, in spite of the rise in Δt.

With a smooth increase in the heat flow the heat-transfer coefficient, after attaining a maximum, drops discontinuously to a value which is characteristic for the pellicular regime of boiling (Fig. 10.2).

The distinguishing feature of this effect is the fact that a restoration to the bubbling regime of boiling (part b) is impossible near that value of the heat flow for which the vapor layer originated. It turns out that an extreme reduction in the magnitude of q is necessary in order to renew boiling in the bubbling regime.

Thus, there exist two critical heat flows: $q_{cr\,1}$, for which the transition from the bubbling regime to the pellicular regime occurs, and $q_{cr\,2}$, for which the vapor film and the restoration of bubbling are destroyed. When the values of the heat flow lie between these critical values, the existence of both regimes of boiling are possible. In such event, depending on the arrangement of the process, the simultaneous coexistence of both regimes of boiling on the same heating surface is possible.

The high heat-transfer intensity during bubbling-type boiling is due to the energetic transfer of the liquid mass from the heating surface in the nucleus of flow as the bubbles grow and burst. The extremely small heat-transfer intensity in the pellicular regime [43] is due to the fact that in this case the liquid mass is separated from the heating surface by a vapor layer having a low thermal conductivity (Fig. 10.3). In Fig. 10.4 and 10.5 photographs of the pellicular boiling of alcohol on a graphite surface are shown; these were obtained by Borishanskii and Tass.

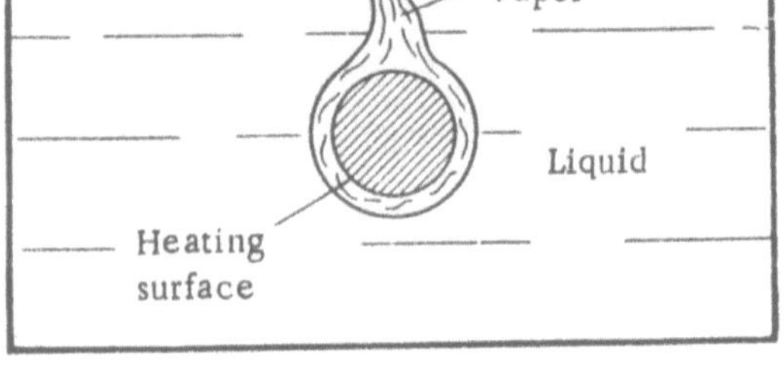

Fig. 10.3. Scheme for pellicular boiling.

Poor wetting of the heating surface by a boiling liquid contributes to the premature origination of the pellicular regime of boiling. In case the liquid does not wet the heating surface at all, stable pellicular boiling is observed for almost any heat flows whatsoever.

The degree of wettability can be characterized by the outer angle between the surface of a body and the vapor bubbles found thereon (or the outer angle between the surface and the drops of a given liquid).

The outer angle is connected with the surface tension of the liquid by the equation

$$\cos\theta = \frac{\sigma_{2,3} - \sigma_{1,3}}{\sigma_{1,2}} - \frac{\psi_{fr}}{\sigma_{1,2}}, \tag{10.1}$$

where σ is the coefficient of surface tension in kg/m and ψ_{fr} is the frictional force in kg/m applied to the wetting perimeter and impeding drops from breaking away.

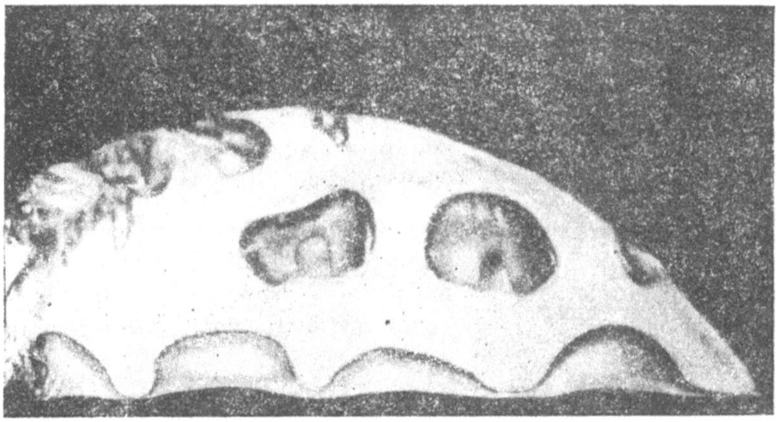

Fig. 10.4. Pellicular boiling of alcohol at a horizontal surface.

The subscripts 1, 2, and 3 refer to the liquid, gaseous, and solid phases, respectively (for example, $\sigma_{1,2}$ indicates the value of the surface tension in a liquid at its boundary with the gaseous phase).

Thus, the quantity θ is determined by the molecular cohesion of the liquid and solid body.

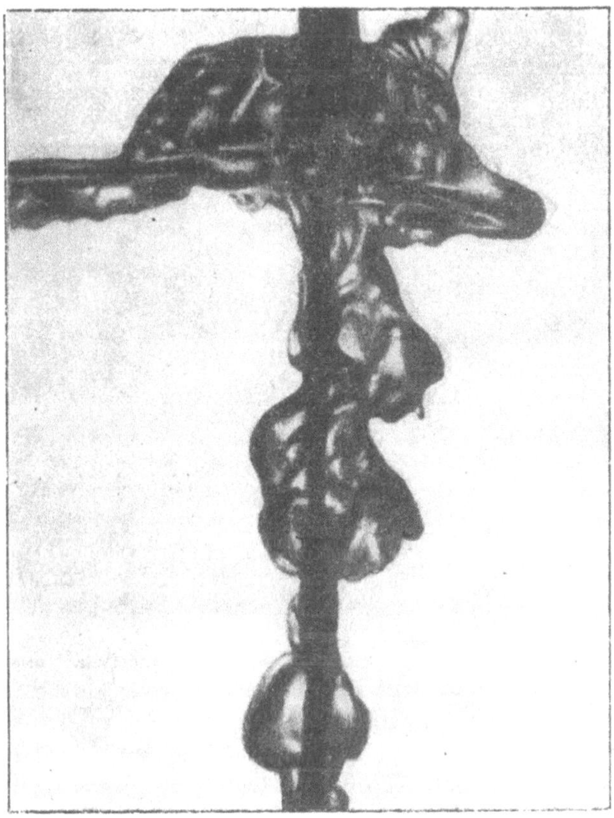

Fig. 10.5. Pellicular boiling of alcohol at a vertical surface.

Acting on the surface layer of a liquid, it is possible, using active substances, to change the degree of wettability of a given surface by a given liquid [34]. Thus, for example, by adding a very small amount of magnesium or sodium to mercury, its contact with steel is effectively improved. The same is observed in the addition of magnesium to a lead-bismuth alloy.

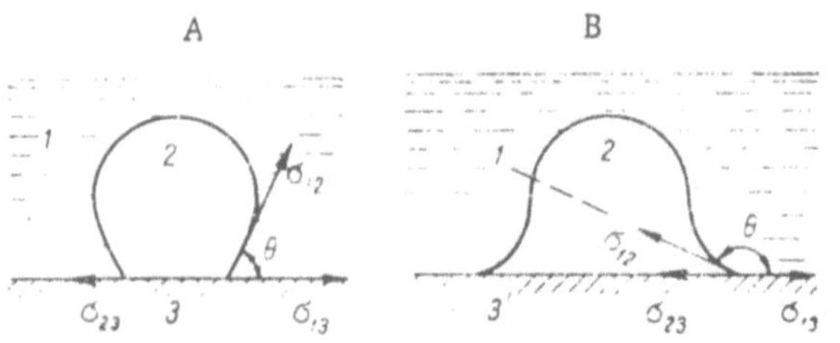

Fig. 10.6. Outer wetting angle θ at a three-phase boundary. A) Steam bubble: 1) water, 2) steam, 3) surface; B) mercury vapor bubble: 1) mercury, 2) mercury vapor, 3) surface.

In Fig. 10.6, the shapes of the vapor bubbles for varying degrees of wetting of the heating surface are shown.

In the preceding chapters it was already noted that many liquid-metal heat-transfer media either do not wet or poorly wet heating surfaces which are made from carbon and stainless steels.

This has quite a strong influence on the heat transfer during boiling, since the degree of wetting of the heating surface proves to have a predominant effect on the nature of the motion of the liquid-vapor mixture.

Effect of Wetting on the Structure and Hydrodynamic Characteristics of a Liquid-Vapor Mixture

In vertical tubes the motion of a two-phase flow, averaged over a very large time interval, always possesses axial symmetry. In inclined and horizontal tubes such a symmetry more or less is maintained only for sufficiently high flow velocities. For small flow velocities the flow becomes stratified: in the lower portion of the tube a liquid is flowing, while in the upper part a vapor (gas) flows. Stratified flow is dangerous in that during the heat exchange the upper part of the tube, which is subjected to vapor flow, does not cool as rapidly as the lower part. This effect is observed both with wetting and with nonwetting of the surface.

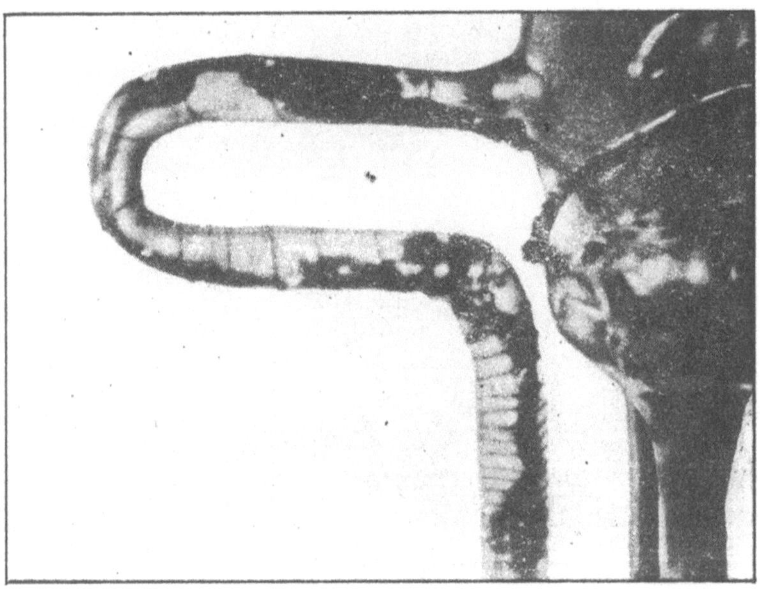

Fig. 10.7. Nature of the motion of a liquid-vapor mixture when the liquid does not wet the heating surface.

In the case of nonstratified and weakly stratified flows the distribution of the liquid and vapor phases essentially depend on the wettability of the tube wall.

A liquid which wets the wall of a tube forms a solid film on it and this insures a high intensity of cooling for the bubbling regime of boiling. In this case, the vapor bubbles which break away from the heating surface become absorbed into the nucleus of flow.

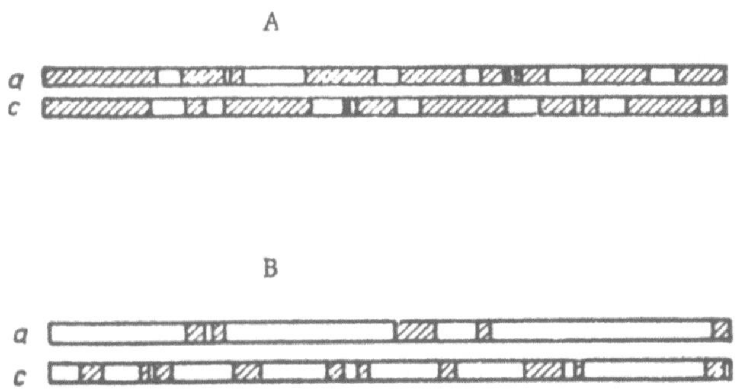

Fig. 10.8. Variation in the conditions of wall contact as a function of the heat flow (cross section in the boiling zone of mercury).
A: $q = 10{,}000$ kcal/m^2•hour, 1.5 sec;
▭) no contact, a) 34%, c) 33.5%;
▨) contact, a) 66%, c) 66.5%;
B: $q = 25{,}000$ kcal/m^2•hour, 1.5 sec;
▭) no contact, a) 87.5%, c) 78.5%;
▨) contact, a) 12.5%, c) 21.5%.

If the liquid does not wet the wall of the tube (for example, mercury in a glass or steel tube), the opposite pattern of motion is achieved, i.e., the vapor bubbles rupture between the wall of the tube and the liquid flow. The liquid, however, moves mainly in the nucleus of flow in the form of a bubbling jet. This pattern of motion is distinctly evident in the photograph (Fig. 10.7).

A graphical-kinematic investigation conducted by Lozhkin and Krol' [35, 3] on the boiling of mercury in a glass tube showed that different points of the heating surface were alternately subjected to the flow of the liquid and vapor in turn. However, the dense adhesion of the mercury to the wall of the tube in this case was not noted. In Fig. 10.8, a composite graph of the variation in the contact conditions as a function of the heat flow is presented. As evident from the figure, with increasing q the fraction of the tube surface contacted by the liquid mercury is sharply decreased.

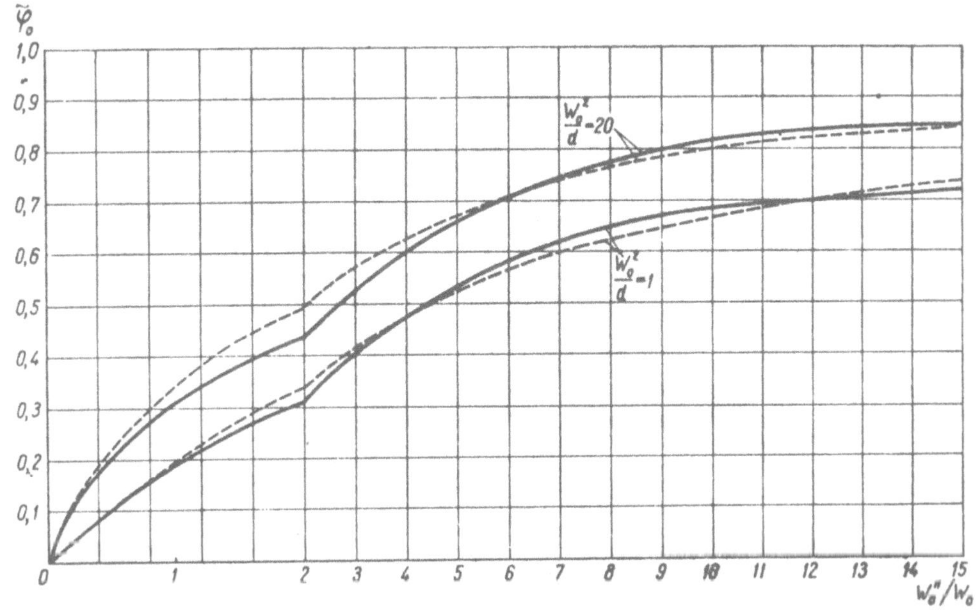

Fig. 10.9. Fundamental hydrodynamic characteristics of two-phase flow for steam-water and vapor-mercury mixtures. ———) Mercury ($p = 5$-15 atmos); – – – – –) water ($p = 5$ atmos).

The variation in the structure of two-phase flow as a function of the wetting of the tube wall by the liquid, appreciably affecting the process of heat exchange, has almost no manifestation in the hydrodynamic resistance. This is connected with the fact that the integral hydrodynamic characteristics of two-phase flow strongly depend on the concentrations by volume of the phases and have little sensitivity to the thin structure of the flow. In this connection, the circulation characteristics of a vapor-mercury mixture turn out to be almost the same as the characteristics of a steam-vapor mixture.

In Fig. 10.9, the fundamental hydrodynamic characteristic of two-phase flow obtained by Gremilov [36] is shown in dimensionless coordinates for vapor-water and vapor-mercury mixtures which are moving in vertical tubes.

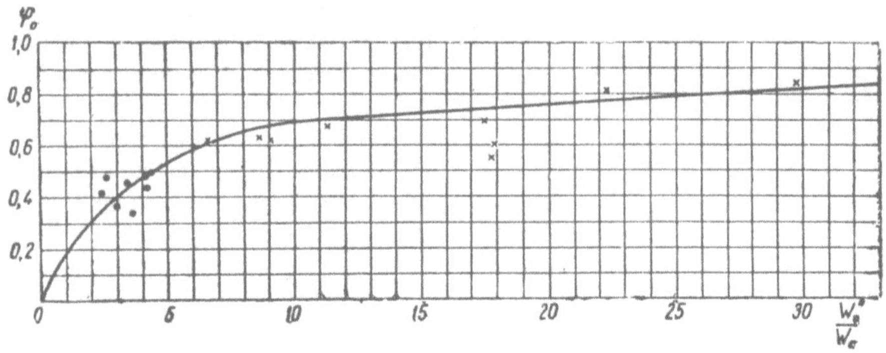

Fig. 10.10. Comparison of the function $\varphi = f\left(\frac{w''_0}{w_0}\right)$ for $\frac{w_0^2}{D} = 1$ with the experimental data. ●) Siryi's tests, x) Korneev's tests.

In Fig. 10.10, the data from the experiments of Siryi [38] with a vapor-mercury mixture are compared w with those of Korneev [37] with a mixture of mercury vapor and a liquid magnesium-mercury amalgam.

In these figures the concentration by volume φ_0 of the vapor in the flow of the mixture is plotted along the vertical axis and the ratio of the reduced velocity of the vapor to the circulation velocity w''_0/w_0 is plotted along the horizontal. The Froude number is assumed as a parameter, with the same accuracy as the quantity $g = 9.81\ \text{m/sec}^2$ and in the form w_0^2/D, where D is the inside diameter of the tube.

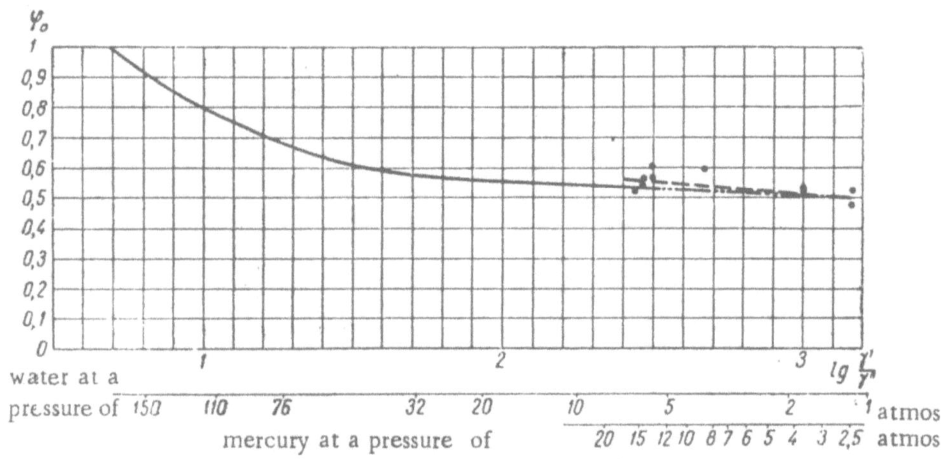

Fig. 10.11. The function $\varphi = f\left(\frac{\gamma''}{\gamma}\right)$ with a constant value of the Froude number and of the ratio of the reduced velocity of the vapor to the circulation rate

$$\left(\frac{w_0^2}{D} = 1,\ \frac{w''_0}{w_0} = 5\right).$$

———) Water; ●) mercury.

In Fig. 10.11, a similar comparison is given for water and mercury.

It is a well-known fact that the parameters just mentioned turn out to have a decisive effect on certain integral hydraulic characteristics of two-phase flow, such as the true gas content φ and the relative pressure drop $\frac{\Delta pD}{\rho' w_0^2 L}$. The effect of the relative density of the phases (γ''/γ), and the viscosity, however, is very small.

Thus, these data once again clearly reveal that the effect of wetting on the integral hydraulic characteristics of two-phase flow is small. Consequently, a calculation of the pressure drop and a calculation of the circulation during boiling of metals in tubes can be effected in the usual way.

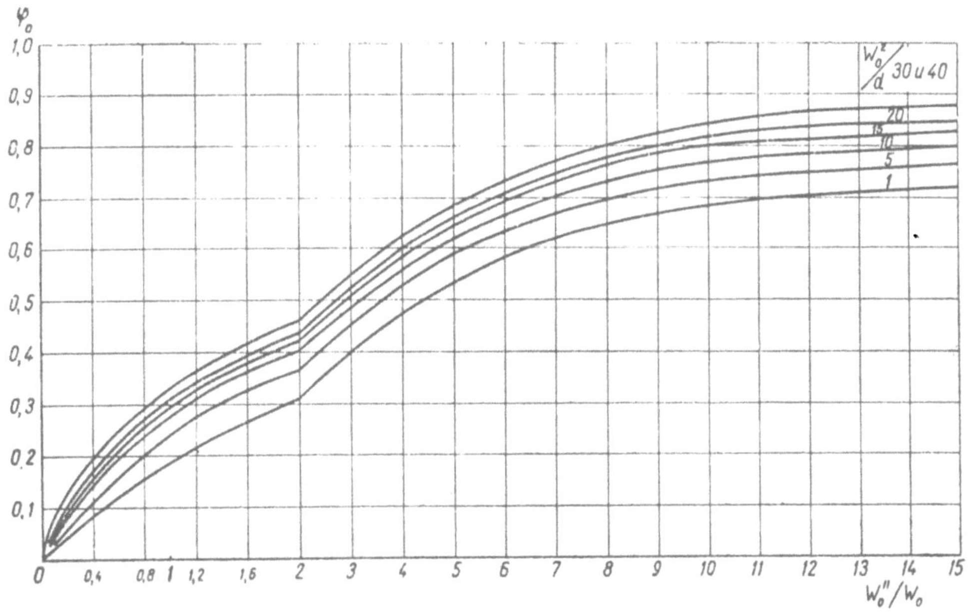

Fig. 10.12. Fraction of the cross section occupied by mercury vapor in a vertical tube, with p = 5-15 atmos (generalized graph).

For mercury and its amalgam the effect of the pressure on the circulation characteristics in the region $5 \leq p \leq 15$ atmos is small enough that they can for all practical purposes be neglected and, for the sake of determining the quantity φ, the single graph cited in Fig. 10.12 can be used.

According to Gremilov's data for inclined tubes, the quantity φ as obtained from Fig. 10.12 should be multiplied by a correction factor, which is determined from Fig. 10.13.

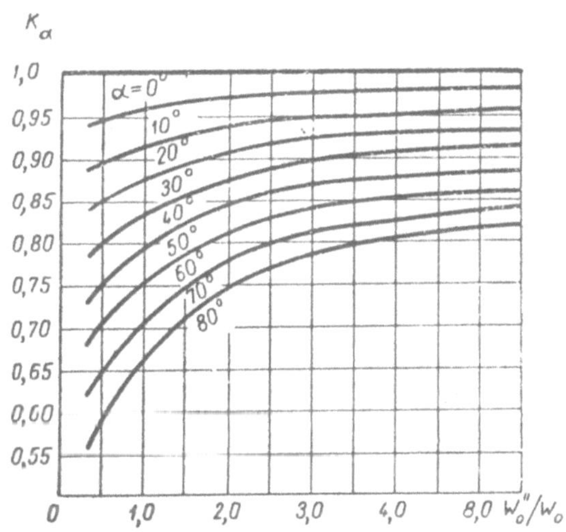

Fig. 10.13. Correction factor for calculating φ in inclined tubes.

The pressure drop can, with an accuracy of ± 15 %, be determined from the well-known equation for a homogeneous mixture

$$\Delta p = \zeta \frac{w_0 \gamma' L}{2gD}\left[1 + \left(1 - \frac{\gamma''}{\gamma}\right)\frac{w_0''}{w_0}\right], \qquad (10.2)$$

where ζ is the coefficient of resistance of a given tube during the flow in this tube of a liquid with $Re = \frac{w_0 D}{\nu'}$.

Effect of Wetting on Heat Transfer During Boiling

It has been shown above that the degree of wetting of a surface by a boiling liquid (including metallic liquids) has a strong influence on the nature of the contact with the heating surface. This fact, which proves to be not too significant for the graphical flow characteristics of a liquid-vapor mixture, has a decisive effect on the heat-transfer intensity in connection with the sharp contrast in the cooling properties of the liquid and vapor phases.

As we have stated previously, a liquid which does not wet the heating surface has little contact with the latter. As a result, a process of heat exchange more nearly approximating the pellicular regime of boiling of wetting liquids and characterized by small heat-transfer coefficients is established. This fact was accurately established by a number of experiments on the boiling of pure mercury on carbon steel heating surfaces in 1936 to 1939; this work was done by Styrikovich, Lozhkin, Kanaev, Kutateladzé, et al. [1-4, 35-37, 40].

Improving the wetting characteristics (for example, by the addition of magnesium or sodium to mercury) leads to a sharp increase in the heat-transfer intensity and to a behavior which is characteristic for the normal bubbling regime of boiling (when $q < q_{cr,1}$); this is particularly evident from Fig. 10.1.

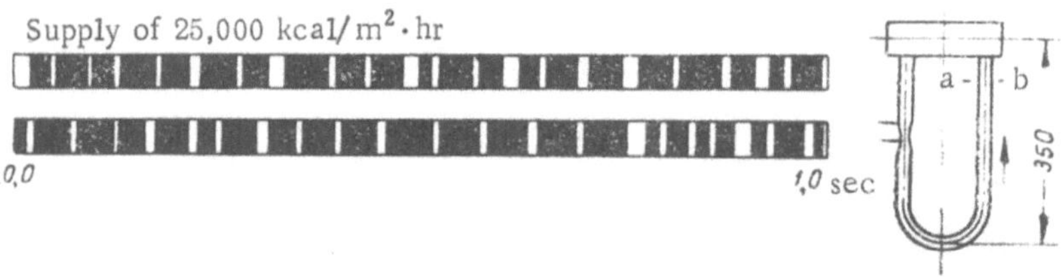

Fig. 10.14. Contact spectrum for the boiling of mercury in a tube with a helix inserted. ▭) No contact, a) 31%, b) 33%; ■) contact, a) 69%, b) 67%.

In addition to the compounding of an amalgam, the heat transfer in the boiling of a liquid which does not wet a given heating surface can be improved to a certain extent by increasing the contact rate.

In Fig. 10.14, a contact spectrum, obtained by Lozhkin [3] for the boiling of mercury in a tube with a flat-iron helix inserted in it, is shown.

As evident from Fig. 10.14, the fractional contact of the heating surface by the liquid in this case attained approximately 70%, while with no baffle (turbulizer) present it amounted to 12-20%.

The corresponding variations in the heat-transfer coefficient are shown in Tables 10.1 and 10.2.

TABLE 10.1
Coefficient of Heat Transfer From A Carbon Steel Tube to Boiling Mercury Without A Mixing Baffle

q, kcal/m²·hr	p, atmos	Number of circulations	α, kcal/m²·hr·°C		
			bottom	middle	top
16 650	1,32	341	1865	729	493
25 400	1,49	219	2460	239	470
29 700	1,57	218	2600	182	300
25 260	1,20	209	2980	1830	349
31 900	1,36	211	5060	414	323
29 700	1,21	194	1990	270	392

In the lower part of the loop there was no boiling (economizer zone), and there rather high values of the heat-transfer coefficient were observed, independently of whether or not the helical baffle was present. In the middle portion of the loop vapor began to form, while in the upper part the boiling process became fully developed. It was especially in these regions that the baffle was most effective.

TABLE 10.2

Coefficient of Heat Transfer From A Carbon Steel Tube to Boiling Mercury With The Insertion of a Mixing Baffle

q, kcal/m²·hr	p, atmos	Number of circulations	α, kcal/m²·hr·°C		
			bottom	middle	top
25 400	1,25	229	1950	2120	5640
48 000	1,10	142	2540	3760	5450
15 100	1,17	134	2400	2890	7780
51 400	1,16	128	2760	2840	6950
55 800	1,14	124	2950	3300	6640
55 800	1,5	132	2490	4130	7060
61 300	1,6	113	3170	8590	8560

Heat Transfer During Boiling In A Large Volume of Liquid

In Fig. 10.15, the dependence of α on q is shown for the boiling of certain liquid metals in a large volume (constructed from the data of Soviet and American authors [6, 45, 46]).

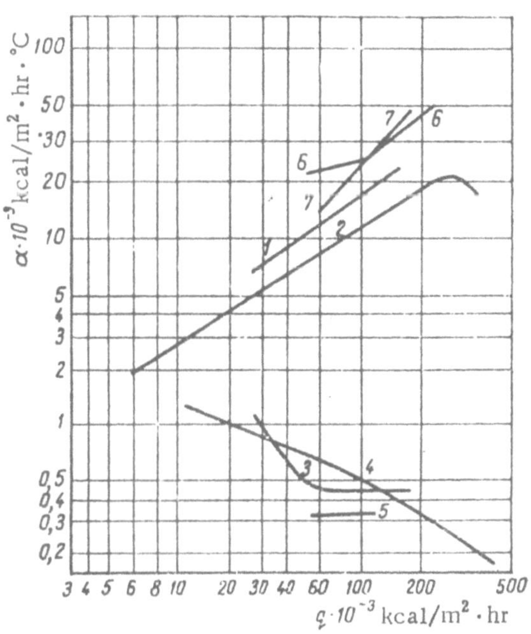

Fig. 10.15. Heat transfer during the boiling of various metals. 1) Boiling of magnesium-mercury amalgam inside a vertical tube; 2) boiling of magnesium-mercury amalgam around a tube in a large volume; 3) pellicular boiling of magnesium-mercury amalgam; 4) boiling of pure mercury; 5) boiling of cadmium; 6) boiling of sodium-potassium alloy; 7) boiling of sodium.

For metallic liquids which wet the heating surface (sodium, sodium-potassium alloy, magnesium amalgam of mercury) in the interval of heat flow $q < q_{cr\,1}$ a normal dependence of the heat-transfer coefficient on the heat flow density is observed:

$$\alpha = Aq^n, \qquad (10.3)$$

where n is approximately 0.6-0.7.

For liquids which do not wet the heating surface (mercury, cadmium) values of α which are characteristic for pellicular boiling are observed.

The addition of a surface-active substance furthers the origination of the bubbling regime of boiling. However, that region of heat flow densities in which this regime is maintained depends on the concentration of the additive.

In Fig. 10.16, Korneev's data [37] on the boiling of a magnesium amalgam of mercury on a vertical steel tube placed in a large volume of the liquid are shown. The experiments were conducted with various concentrations of magnesium. As evident from the figure, for different concentrations of magnesium the function $\alpha(q)$ in the bubbling region is almost one and the same. However, the transition to pellicular boiling takes place for different heat flow densities. This becomes clear in Fig. 10.17, where the dependence of the quantity $q_{cr,\,1}$ for mercury, boiling at atmospheric pressure, on the concentration by weight of the magnesium is shown.

An increase in the concentration of the amalgamating substance increases the adhesion of the particles of the liquid-metal heat-transfer medium to the heating surface and insures boiling in the bubbling regime over a large range of q.

The effect of the tube diameter during bubbling-type boiling of a metal is practically nonexistent (verified for D up to 40 mm), just as in the case of nonmetallic liquids.

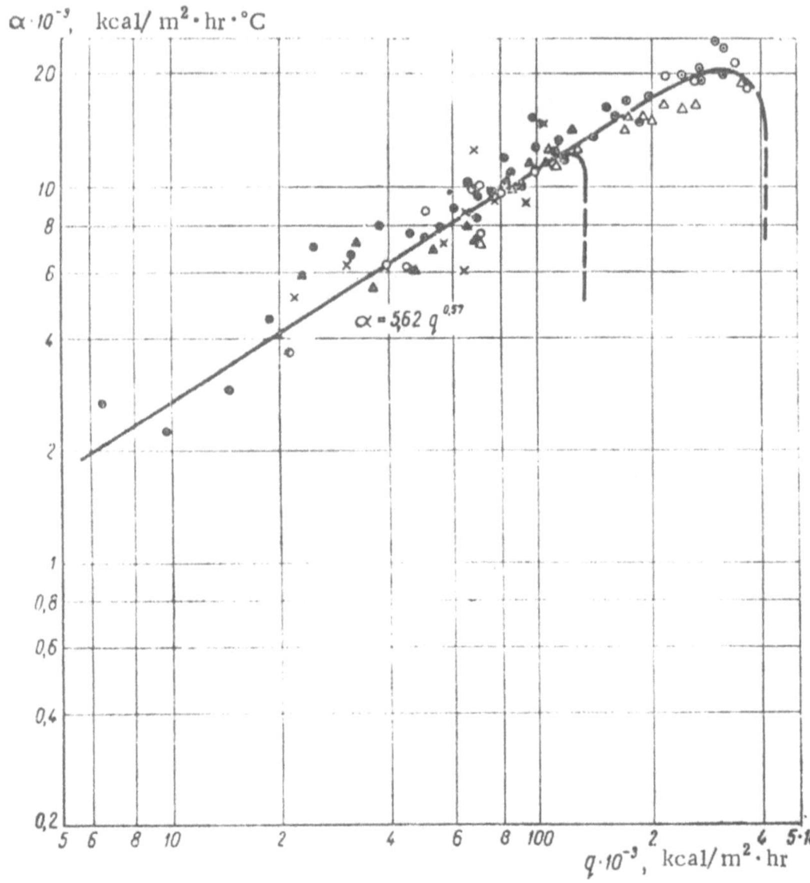

Fig. 10.16. Heat transfer during the boiling of magnesium-mercury amalgams of various concentrations (tube placed vertically in a large volume, D = 21.5 mm). The different points designate the values of the magnesium concentration ξ, ranging from 0.01 to 0.03%.

The data presented in Fig. 10.15 for boiling in the bubbling regime of sodium and a magnesium amalgam of mercury can be compared with calculations on the basis of the following equations as criteria, obtained from experiments with nonmetallic liquids.

The equation of Averin and Kruzhilin [39]:

$$\frac{\alpha}{\lambda}\left(\frac{\sigma}{\gamma-\gamma''}\right)^{1/2}=0{,}082\left(\frac{\nu}{a}\right)^{0,45}\left[\frac{\gamma'' r q}{\Delta T_S \lambda(\gamma'-\gamma'')}\right]^{0,7}\times$$
$$\times\left[\frac{c\gamma^{0,5}\Delta T_S\,\gamma\,(\gamma-\gamma'')^{0,5}}{(r\gamma'')^2}\right]^{0,333} \tag{10.4}$$

The equation of Kutateladze [40]:

$$\frac{\alpha}{\lambda}\left(\frac{\sigma}{\gamma-\gamma''}\right)^{0,5}=0{,}44\left(\frac{\nu}{a}\right)^{0,35}\left(\frac{qp\cdot 10^{-4}}{r\gamma''(\gamma-\gamma'')\,\nu}\right)^{0,7}. \tag{10.5}$$

Borishanskii and Minchenko, working on different experiments [41], came to the conclusion that for ordinary liquids, in equation (10.5) it would be possible to assume a power exponent equal to 0.7 for the Pr number, while the coefficient of proportionality should be equal to 0.55. Then

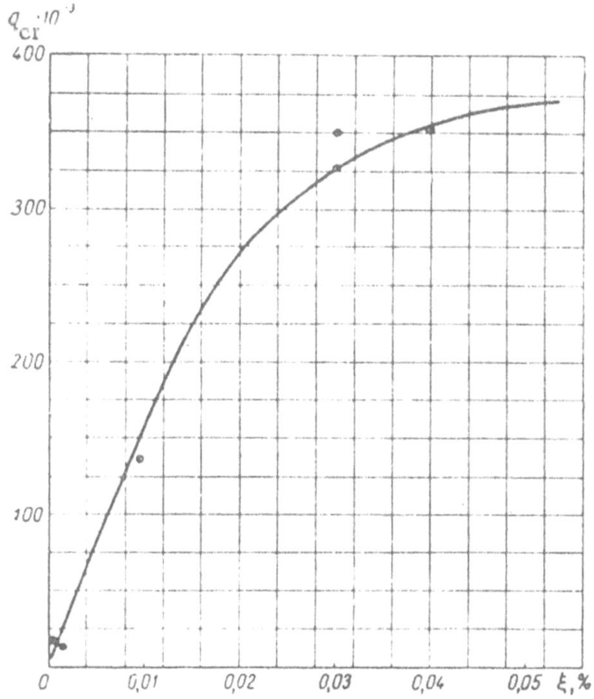

Fig. 10.17. Effect of magnesium concentration in mercury on the critical heat supply for boiling in a large volume.

$$\frac{\alpha}{\lambda}\left(\frac{\sigma}{\gamma-\gamma''}\right)^{0,5}=$$

$$= 0,55\left(\frac{qp\cdot 10^{-4}}{r\gamma''(\gamma-\gamma'')a}\right)^{0,7}. \qquad (10.6)$$

This equation lends support to the assumption that, owing to the intense turbulization, molecular friction in the boiling boundary layer can be neglected [40].

A criterion, for which the relation (10.6) is a special case, has the form

$$\frac{\alpha}{\lambda}\sqrt{\frac{\sigma}{\gamma-\gamma''}} = \text{const}\left[\frac{p}{\sqrt{\sigma(\gamma-\gamma'')}}\right]^m \times$$

$$\times\left[\frac{q}{r\gamma'' a}\sqrt{\frac{\sigma}{\gamma-\gamma''}}\right]^n. \qquad (10.7)$$

In equation (10.6) it is assumed that $m = n = 0.7$.

Heat Transfer During Boiling In A Tube

It is known that in fully developed boiling of the bubbling type the heat-transfer coefficient very slightly depends on the flow velocity of the liquid. Korneev's data, cited in Fig. 10.18, on heat-transfer during the boiling of a magnesium amalgam at the surface of a tube under the conditions of free convection and inside a vertical tube with natural circulation are mutually similar, i.e., they support this rule for metals also.

TABLE 10.3
Comparison of the Experimental Values of the Coefficient A in Equation (10.3) With Calculated Values

Substance	Pressure p, atmos.	$A = \alpha/q^{0,7}$			
		experimental (average values	according to Equation (10.4)	according to Equation (10.5)	according to Equation (10.6)
Magnesium-mercury Amalgam.	1,03	3,30 ($q<2\cdot10^5$)	24,6	5,19	1,1
The same	10,3	4,75	27,5	6,42	1,5
Sodium	1,03	($q<1\cdot10^5$) ~8	56,6	22,6	3,72

Note. As evident from the table, the order of magnitude of α for the mercury amalgam fits the experiments best for the calculation according to equation (10.5). For sodium equations (10.5) and (10.6) give approximately equal discrepancies.

There only remains the situation when the liquid does not wet the wall of the tube.

In Fig. 10.19, the results of the experiments of Styrikovich, Semenovker, and Sorin [1] on heat transfer during the boiling of mercury in vertical steel tubes are shown. The significant effect of the circulation velocity w_0 and the tube diameter D are clearly apparent. A certain decrease in the magnitude of α with increasing q is also noted. The effect of pressure is slight.

In Fig. 10.20, are shown the reduced experimental data obtained by the authors of [1].

From this figure the order of magnitude of α for the boiling of mercury in a vertical tube which is not wetted by it can be estimated.

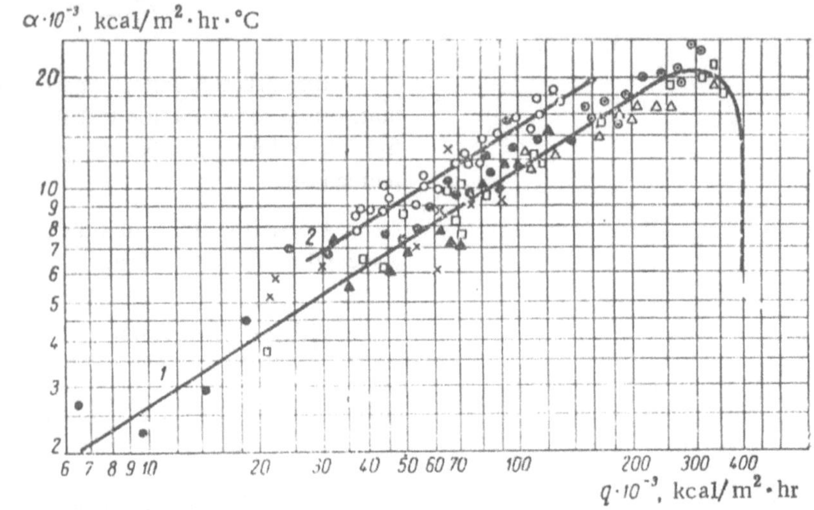

Fig. 10.18. Comparison of experimental data on heat transfer during the boiling of a magnesium-mercury amalgam. The various symbols refer to different magnesium concentrations in the mercury, ξ = 0.01-0.04%; 1) large volume; 2) tube (natural circulation).

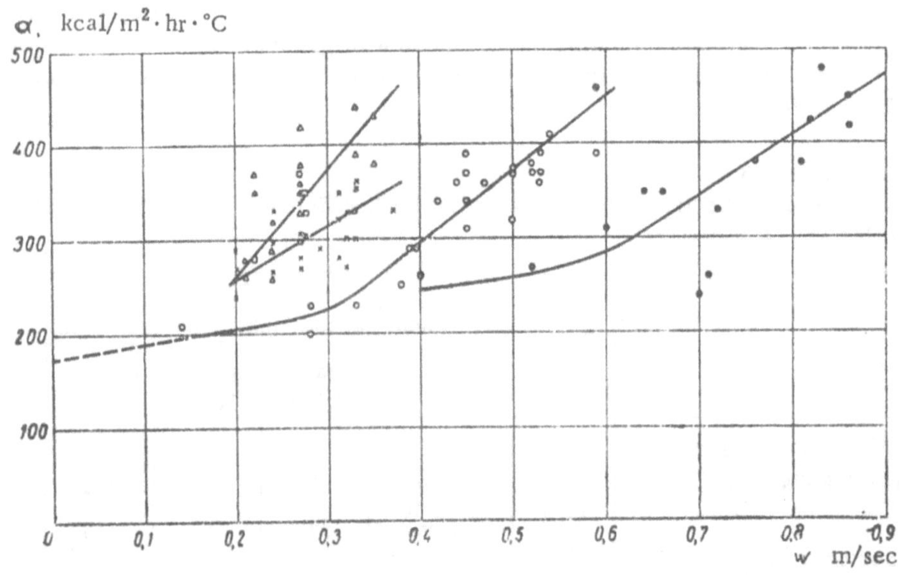

Fig. 10.19. Heat transfer to boiling mercury flowing in vertical tubes with various tube diameters and heat loads.

Symbol	D, mm	q, kcal/m²·hr
●	21	25 000
○	32	25 000
□	40	25 000
△	40	50 000
×	40	70 000

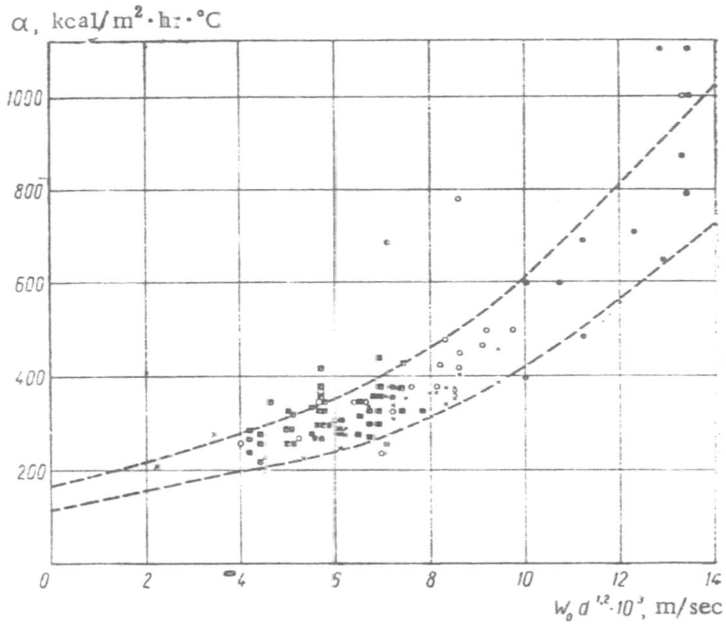

Fig. 10.20. Composite graph on the heat transfer to boiling mercury in tubes.

Symbol	D, mm	q, kcal/m²·hr	Tube orientation
○	21	25 000	vertical
⊗	20	33 000—40 000	inclined
◐	22	35 000—69 000	the same
●	20	91 000—98 000	"
×	32	25 000	vertical
□	40	25 000	the same
◩	40	50 000	"
■	40	70 000	"

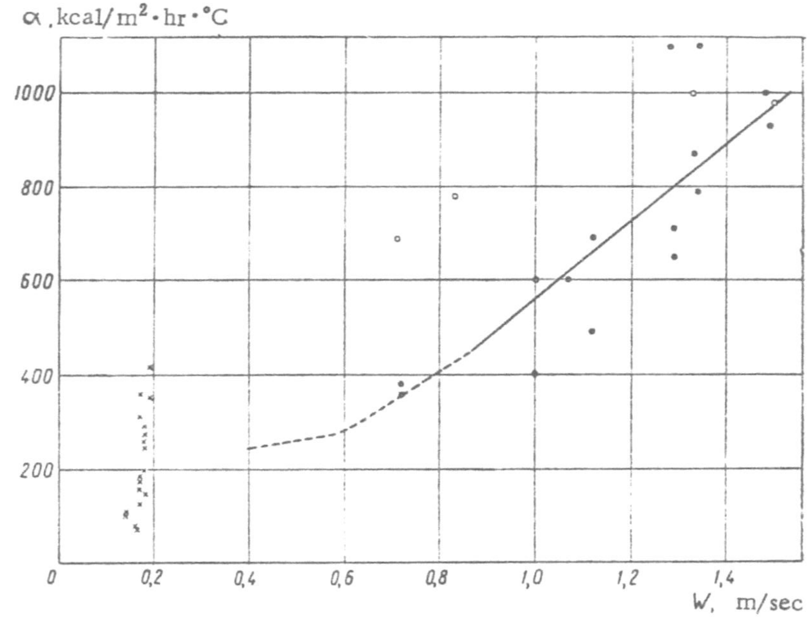

Fig. 10.21. Heat transfer to boiling mercury moving in a nonstratified flow in an inclined tube.

Symbol	D, mm	q, kcal/m²·hr
○	20	33 000—40 000
◐	20	55 000—69 000
●	20	91 000—98 000
– – – –	21	Vertical tubes
×	22	Experiments of Lozhkin and Kanaev

In Figs. 10.21 and 10.22 the results of experiments by the same authors are shown for inclined tubes. It is evident that in nonstratified motion the heat exchange in inclined and vertical tubes is practically identical. In the case of stratified motion the heat exchange is sharply diminished in the upper portion of the tube and is also reduced, but to a lesser degree, in the lower part.

The lessening of the heat exchange does not set in at the boiling point of the entire mass of the liquid, when the latter does not wet the wall of the tube, but rather at an earlier time, soon after the temperature of the inner wall reaches the saturation temperature at the pressure existing at a given cross section.

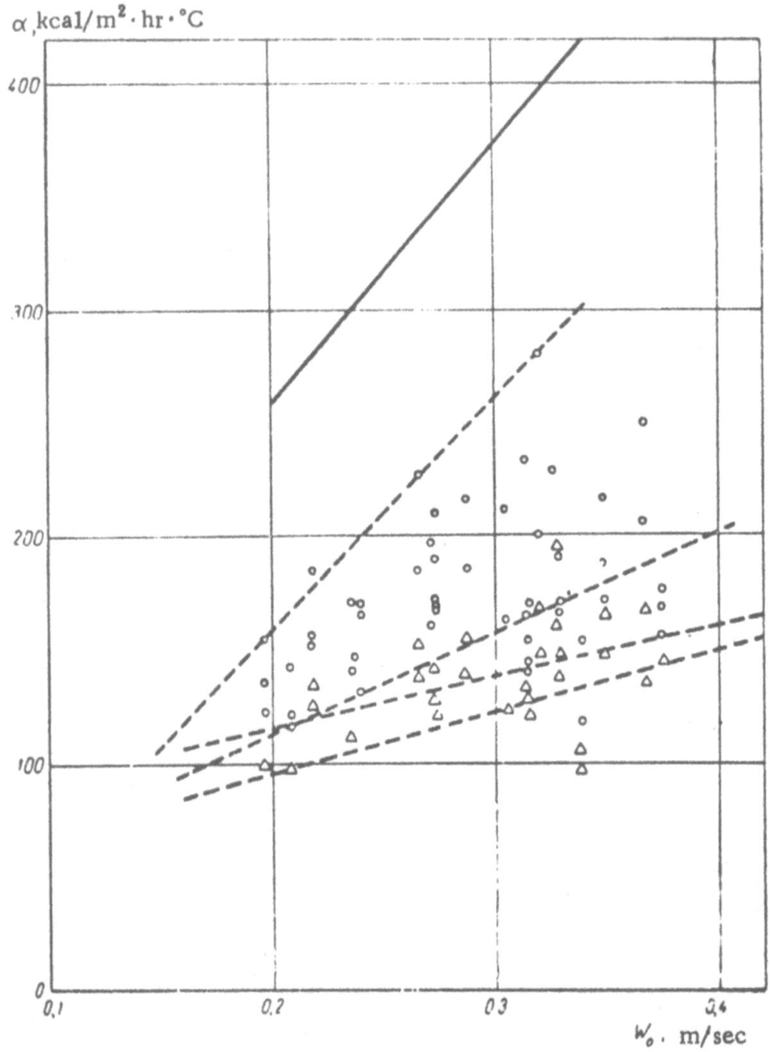

Fig. 10.22. Heat transfer to boiling mercury moving in a stratified flow in an inclined tube. O) Lower generatrix; Δ) upper generatrix; ——) straight line obtained for vertical tubes (D = 40 mm).

Experiments on heat transfer during the boiling of a magnesium amalgam of mercury inside horizontal tubes were conducted by Korneev [14, 37].

In Fig. 10.23, the dependence of the heat-transfer coefficient on the reduced velocity of the liquid phase w_0' is shown, assuming some fixed value of the reduced velocity of the vapor. The heat-transfer coefficient is determined independently from measurements of the temperature at the upper and lower generatrices of the tube.

As evident from the figure, the reduction in the heat transfer in the upper portion of the tube depends on the heat flow density, which determines the increase in the vapor content in the upper portion of the nonsymmetric two-phase flow. The velocity $w'_{0,1}$, above which the heat transfer in the upper portion of the tube remains at an almost constant high level, becomes greater as q becomes higher, and can be determined from the empirical equation

$$w'_1 \approx 22 \cdot 10^{-5} q^{0.42} D^{0.76} \text{ m/sec,} \tag{10.8}$$

where q is in kcal/m²·hour and D is in mm.

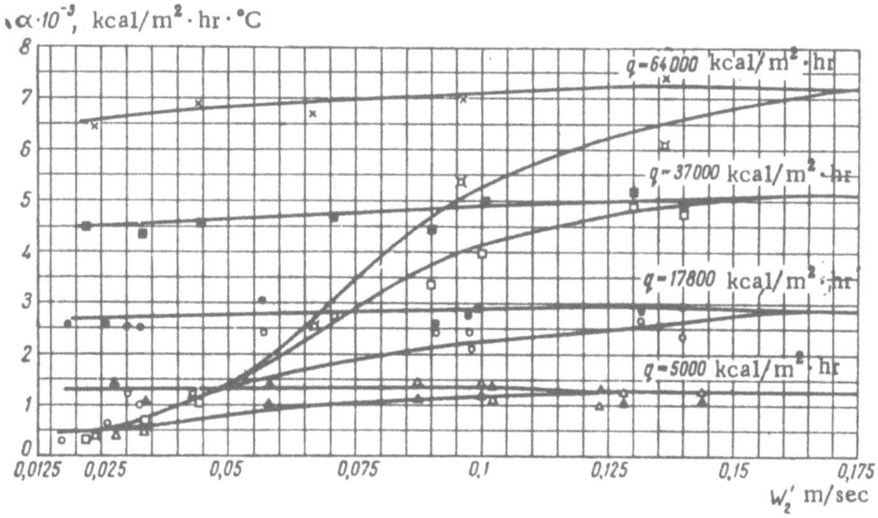

Fig. 10.23. Dependence of heat transfer from the wall to a magnesium amalgam on the reduced velocity of the liquid w'_0 and the heat supply q for a horizontally heated tube (D = 17.6 mm) with a vapor velocity $w'' = 2$– m/sec.
×) Bottom, ¤) top = 64,000 kcal/m²·hour, ●) bottom, ○) top = 18,000 kcal/m²·hour,
■) bottom, □) top = 31,000 kcal/m²·hour, ▲) bottom, △) top = 5000 kcal/m²·hour.

When $w'_0 > w''_{0,1}$ the heat-transfer coefficient was determined by the empirical equation

$$\alpha = 12 q^{0.67} w_0^{0.3} D^{-0.45}. \tag{10.9}$$

The equations are applied for $5000 < q \ll 70{,}000$; $13 < D \lesssim 40$, $1 < p \leqslant 12$ atmos., $1 < w''_0 \leq 19$ m/sec.

Chapter 11

HEAT-EXCHANGING EQUIPMENT

Some Design Features of Liquid-Metal Heat-Exchangers

Heat-exchanging equipment is used in nuclear power plants for the transmission of heat to a working substance, which is delivered to a turbine. Such a working substance can be either steam or a gas. The general plan of the secondary loop and its design elements are chosen with consideration for the whole complex of engineering problems connected with the cooling of a reactor and the insurance of safe operation.

The transfer of heat from the heat-transfer medium of the primary loop can be effected either directly to the gas or water (through a heat exchanger) or through a second (intermediate) liquid-metal heat-transfer medium. In the latter case the power arrangement as a whole is a three-loop affair.

In the application of an intermediate loop the pressure in it is necessarily specified at a somewhat higher value than in the primary loop. In this way, in case of overflow of the heat-transfer medium the secondary loop will be protected against radioactive contamination.

In steam generators in which a sodium-potassium alloy serves as the heat-exchange medium, particular attention must be given to the elimination of direct contact of the first and second heat-transfer media. One step is the construction of an intermediate layer between the heat-transfer medium and the water, for example, a layer of mercury or other metal, which is inert with respect to water and sodium (or sodium-potassium alloy). Such an intermediate filler should have a sufficiently high heat conductivity, and especially since the weight of the filler in the vapor generator can be considerable, it should be inexpensive.

In liquid-metal heat-exchangers ordinary temperature-expanding fittings and devices are inapplicable. The hermeticity of such systems can be insured only by the highest quality welding with the application of highly sensitive flow-meters. The temperature compensation of heat-exchanger tubes should be insured by the shape of the heating surface, i.e., by the presence of appropriate bends. In connection with the high constant-volume heat capacity of the working substance and the high heat-transfer coefficients the diameters of the tubes are usually made small. Good results in the compensation for temperature expansion are given by the application of welded dynamophones. Both single and binary dynamophone seals are applied. In the latter case the space between the inner and outer dynamophone is filled with a neutral gas.

In steam generators the heating surface is usually divided into three parts: the economizer, in which water is heated to the boiling point, the vaporizer, in which the process of steam formation takes place, and the superheater, in which the steam is superheated. Between the vaporizer zone and the steam-superheater zone a steam-separation apparatus is allowed for.

There are two types of steam generator: the direct flow and naturally-circulating drum types. The maximum temperature of heating is determined by the temperature of the liquid metal as it enters the steam generator. Consequently, such equipment, as a rule, should be built around the principle of counterflow. The free expansion of the heat-exchanger tubes can be insured, for example, by constructing them in the shape of right-angle or U-tubes.

The Protective Layer

As already pointed out above, for reducing the probability of a reaction between the liquid-metal heat-transfer medium and the water in the steam generator to a minimum, a protective layer of some medium which is neutral with respect to both heat-transfer media could be used.

The main shortcomings of such a construction are the following: the complexity of the allocation of the heat-exchange surfaces and their preparation, the reduction in heat transfer due to the high thermal resistance of the protective layer, and the high costs of the equipment.

Two basic variants in the design of the protective layer are possible: 1) Double-walled tube constructions consist of inner tubes, through which the medium with the higher pressure (in steam generators, water, and steam) flows, and an outer tube. The protective medium fills the gap between these tubes. The second heat-transfer medium flows in the space between the tubes which is formed by the outer tubes and the body of the heat-exchanging apparatus. 2) Constructions with independent tube systems consist of the casing, in which rows of tubes with the primary heat-transfer medium flowing in them are placed alternately with those carrying the secondary. The protective medium in this case fills the space between the tubes.

By inserting various indicators (for example, pressure indicators) into the space filled with the protective medium, it is possible to record the accidental occurrence of leakage of the heat-transfer medium. Thus, if we maintain the pressure in the protective layer at a lower level than that of the vapor, but higher than the pressure of the liquid-metal heat-transfer medium, then the seepage of the steam through a leak in the walls of the tube will raise the pressure in the protective space, while leakage into the tubes from the direction of the liquid metal will lower this pressure.

The thermal resistance of the interlayer per unit length (in meters) of tube, when the presence of convection effects is not significant, is determined by the usual relation:

$$R^* = \frac{1}{2\pi\lambda} \ln \frac{R_2}{R_1}, \qquad (11.1)$$

where λ is the thermal conductivity of the protective layer, and R_1 and R_2 are the outer and inner radii of the annular gap.

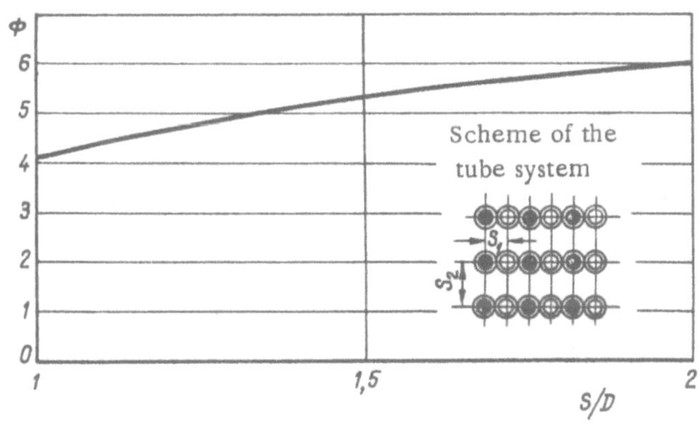

Fig. 11.1. Dependence of the "form factor" Φ on the geometric parameters of the system.

For two parallel infinite rows of tubes the thermal resistance of the space between the tubes, when no convection is present, is a function of the relative geometric dimensions of the system and the thermal conductivity of the protective medium. If the temperature of the outer surfaces of the tubes of the heat-delivering and heat-absorbing rows are t_1 and t_2, respectively, then the heat flow from one meter of length of the half-surface of one tube is equal to

$$Q = \frac{t_1 - t_2}{R^*} \lambda\Phi (t_1 - t_2), \qquad (11.2)$$

where Φ is the so-called "form factor" of the system.

The nature of the dependence of Φ (obtained by Krzhizhanovskii) on the geometric parameter S_1/D is shown in Fig. 11.1.

Heat Transfer

The heat transfer is ordinarily determined by the equation

$$Q = kF\Delta\bar{t}, \tag{11.3}$$

where Q is the quantity of heat delivered (the heating power of the apparatus, in kcal/hr), F is the heat-exchange surface area in m^2, and Δt is the mean temperature difference in ° C.

When in one casing surfaces with very different heat-transfer conditions are present (for example, the vaporizer and superheating zones in steam generators), calculations according to Equation (11.3) are performed separately for each part of the apparatus.

The magnitude of the heat-transfer coefficient k is usually referred to the mean diameter of the tubes which form the heating surface, and is calculated from the equation

$$k = \frac{1}{\frac{1}{\alpha_1} + R^* + \frac{1}{\alpha_2}}, \tag{11.4}$$

where α_1 and α_2 are the heat-transfer coefficients corresponding to the primary and secondary heat-transfer media, in kcal/m^2·hour·°C, and R^* is the thermal resistance of the walls and the protective layer.

The application of different types of partitions, serving as fortification of the tube system and as a lengthening of the path travelled by the heat-transfer medium, in heat-exchanging equipment complicate the choice of analytic scheme for determining the magnitudes of α_1 and α_2 with the aid of the equations obtained in studying the heat exchange under the ideal conditions described above. In this case we resort to rules which have been developed over many years of calculations in heat-exchanger practice (see, for example the norms for calculating a boiler aggregate [59]).

The results of direct thermal testing of liquid-metal heat-exchangers as yet are only a few. The data given below make it possible to arrive at some judgment on the order of magnitude of the heat-transfer coefficients of such arrangements. However, it should be borne in mind that these quantities are closely connected with the design features of the heat-exchanging surfaces and their absolute dimensions.

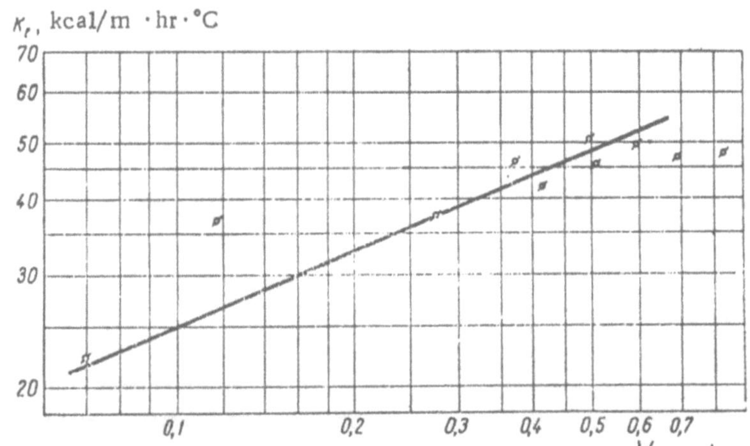

Fig. 11.2. Heat transfer in a heat-exchanger, assuming longitudinal flow along the tubes in the space between them.

In Fig. 11.2, the dependence of the coefficient of heat transfer from a heavy metal to sodium in an experimental heat-exchanger tube is shown. The sodium is flowing inside the tubes, while the heavy metal is flowing along the tubes in the space between them.

Qualitatively, similar results were obtained by Brooks and Rosenblatt [25-27] in testing heat-exchangers with sodium and sodium-potassium alloys as heat-transfer media and vaporizers with a sodium-potassium alloy as the primary heat-transfer medium. The diameters of the tubes were 10-16 mm, the length of the heat-exchange zone was of the order of 1-2 m.

In Fig. 11.3, the generalized data presented by these authors on the heat-transfer coefficient for a liquid metal flowing in the space between the tubes are shown. The heat-transfer coefficient was not determined from measurements of the wall temperature (such measurements were not carried out), but by analytic means from the experimental values of the heat transmission and the calculated value of the coefficient of heat transfer from the inside of the tube. An average logarithmic straight line curve, drawn through this point, is described by the equation

$$Nu = 61\, Pe^{0,6} \left(\frac{\Omega}{F}\right)^{1,2}, \tag{11.5}$$

where the outside diameter of the tubes is used in the Nu and Pe numbers, while the ratio of the area of the effective cross section of the space between the tubes Ω to the heat-exchange surface F is analogous to the ratio D/L.

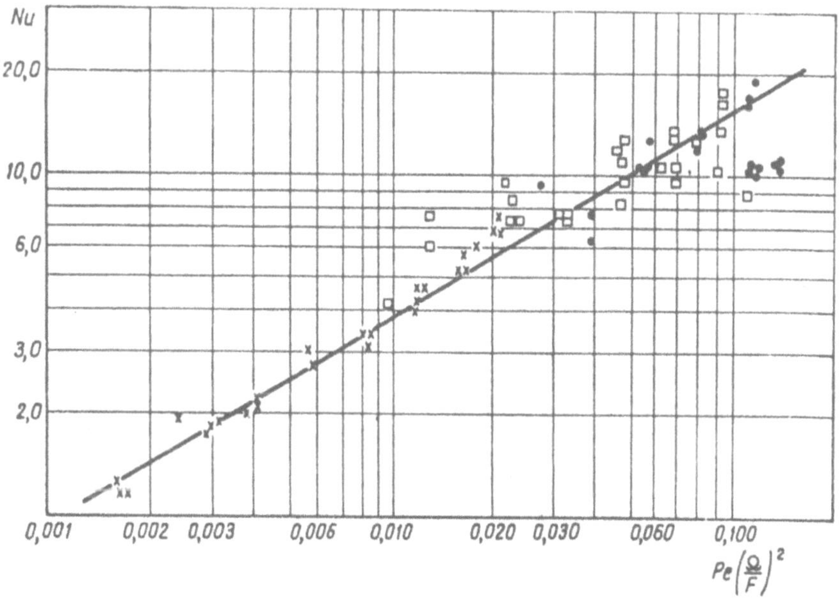

Fig. 11.3. Generalized experimental data on the heat transfer during the flow of a liquid metal in the space between the tubes. x) Heat-exchanger; ●) naturally-circulating steam generator; □) forced-circulation steam generator.

The structure of equation (11.5) is almost the same as the structure of the equation for heat transfer during longitudinal flow around a plate. In fact, if we apply the equations of Chapter 6 to the case of flow along tubes by an unbounded liquid metal stream, then we obtain the following:

a) for a laminar boundary layer according to equation (6.16)

$$\frac{\bar{\alpha}}{\lambda} D \sim \left(\frac{w_0 D}{a}\right)^{0,5} \left(\frac{D}{L}\right)^{0,5}; \tag{11.6}$$

b) for a turbulent boundary layer according to equation (6.43)

$$\frac{\bar{\alpha}}{\lambda} D \sim \left(\frac{w_0 D}{a}\right)^{0,65} \left(\frac{D}{L}\right)^{0,85}. \tag{11.7}$$

The power exponent associated with the Pe number in equation (11.5) is very near the power exponent in equations (11.6) and (11.7). The relative length of the heat-exchanging apparatus turns out to have a much greater effect. In this respect, the experimental results, as expressed by equation (11.5), apparently have a limited value. A more detailed investigation of the same experimental data shows that they do not exhibit quite the same behavior for different types of testing equipment. This is evident from Fig. 11.4, where a common line is not drawn through the points, but instead two lines, corresponding to the heat-exchanger and to the steam generators.

The influence of the construction and arrangement of the apparatus on the heat transmission is expressed also in the experimental data of Brooks and Rosenblatt on the general heat-transfer coefficient (Fig. 11.5).

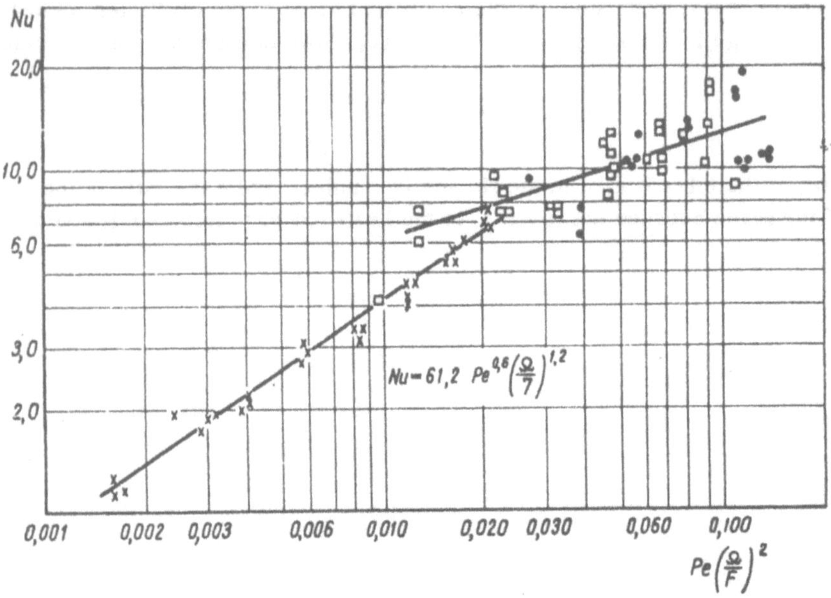

Fig. 11.4. Generalized experimental data on heat transfer during the flow of a liquid metal in the space between the tubes. x) Heat-exchanger; ●) naturally-circulating steam-generator; □) forced-circulation steam-generator.

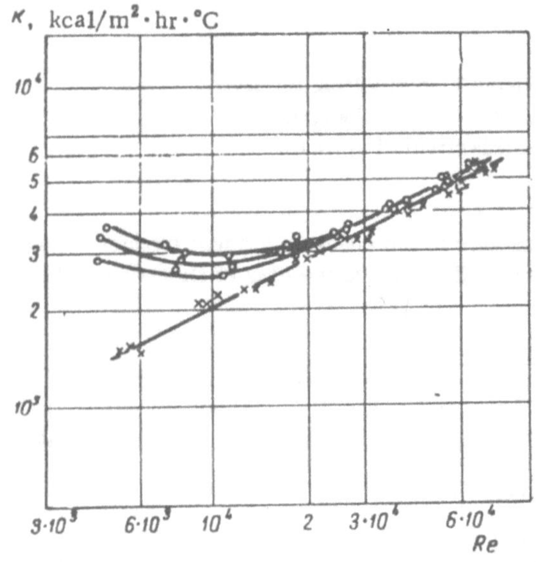

Fig. 11.5. Effect of the construction and arrangement of the apparatus on heat transmission. O) Vertical apparatus; x) horizontal apparatus.

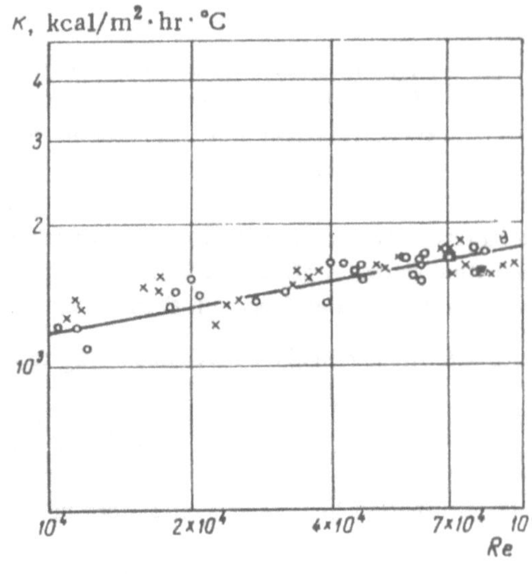

Fig. 11.6. Heat transmission in a vaporizer with sodium-potassium as the primary heat-transfer medium. O) Heat-exchanger No. 2; x) heat-exchanger No. 3.

In Fig. 11.6, data on the heat-transfer coefficient for a sodium-potassium alloy as a heat-transfer medium in a vaporizer are shown.

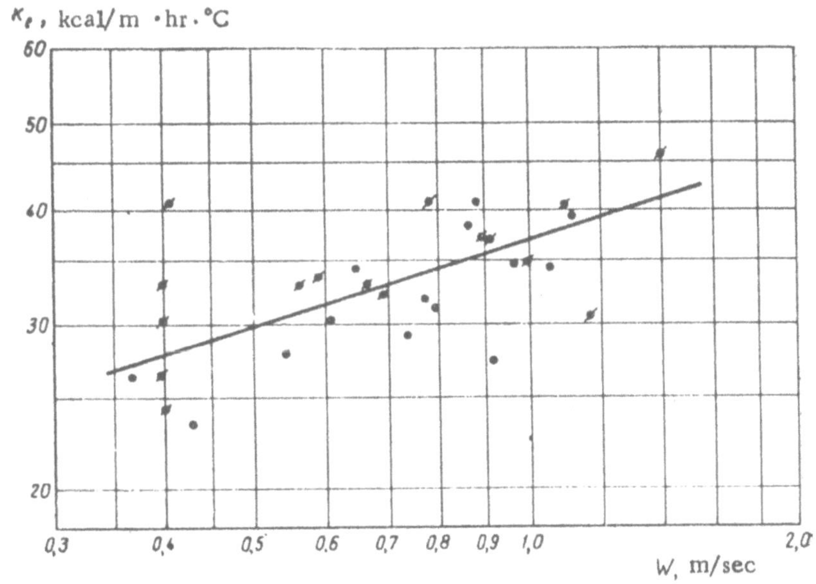

Fig. 11.7. Heat transmission in a sodium-potassium heat exchanger. ●) Heat-exchanger No. 2; ⌀) heat-exchanger No. 3.

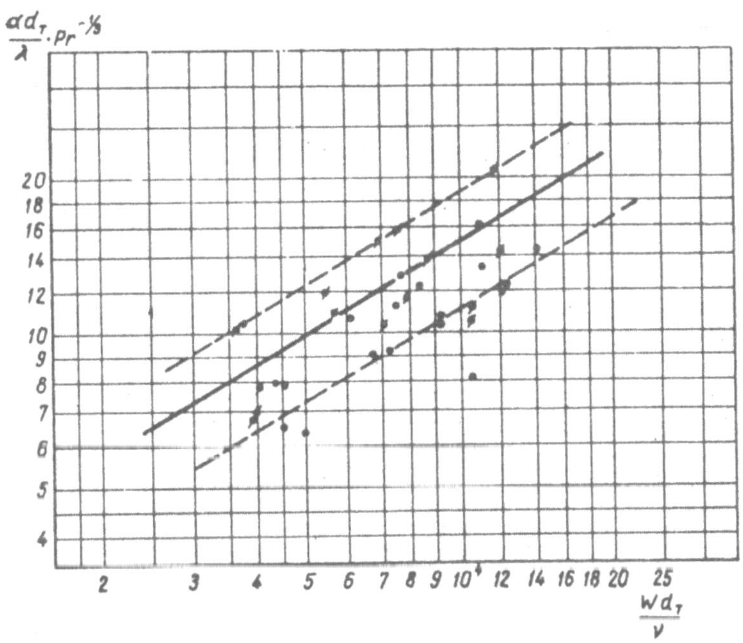

Fig. 11.8. Heat transfer in the space between the tubes. ●) Heat-exchanger No. 2; ⌀) heat-exchanger No. 3; - - - - -) limiting maximum slope of the experimental points.

In Figs. 11.7 and 11.8, are shown the experimental results of Tidball [24] with a tubular heat-exchanger. The tube diameter was equal to 2 mm, the length of the heat-exchange zone was equal to 200 mm. In the space between the tubes there were seven transverse partitions, which caused mixed (lateral-longitudinal) flow around the inner surfaces of the tubes.

The heat-transfer coefficient for the space between the tubes was calculated from an experimental value of the heat transmission and the calculated value of the heat-transfer coefficient inside the tube.

Chapter 12

STABILITY OF HEAT-RESISTANT MATERIALS IN LIQUID METALS

Attack by Liquid Metals on Construction Materials

Solids which are in contact with liquid metals are subjected to corrosive and erosive destruction. Erosion arises under the action of a liquid metal mainly near branch points in the flow and in regions where the boundary layer is disrupted. The subsequent erosion of each new layer of the surface of the solid furthers the development of corrosion. Corrosion in turn weakens the structure of the surface layer and intensifies erosive destruction.

Tubes and other compounds subjected to smooth flow do not disclose any marked erosive destruction effects even after long exposure to attack by the medium, flowing with very high velocities.

One type of interaction between the construction metal and the liquid-metal heat-transfer medium is the possible formation of a solid solution. The solid solution can be of either the interstitial or substitutional types [60, 69]. The formation of a solid solution takes place whenever a component having a large atomic radius is one of the transition metals, while the ratio of the diameter of the interstitial atom to that of the solvent atom is less than 0.59.

The formation of a solid solution by substitution takes place when the following conditions are observed: 1) the radii of the atom should not differ by more than 15% for the formation of solid solutions with a wide range of solubility; 2) the metals should have similar crystal structures; 3) the metals should have similar rows in the stress tensor.

The solubility is greater in those metals for which the melting point is lower.

In steels and other complex alloys the interaction of the liquid metal and the admixtures and dissolved substances contained in it with the separate components of the construction material and the intercrystalline corrosion of the latter are of considerable significance.

The concentration of a well dissolved component of an alloy in a construction material turns out to have a considerable effect on its leaching-out intensity. For example, in steels with a nickel base which are placed in liquid bismuth or lead, the depletion of the nickel proceeds relatively weakly when its content in the steel is less than 20-25% and very intensely when the content is greater than 28-30%.

In the alkaline metals "self-welding" on the basis of diffusion bonding clearly appears [61, 62]. Thus, if two pieces of a solid metal in contact are immersed in sodium at a temperature of the order of 500° C and higher, then after a certain amount of time both pieces will appear to be tightly bonded with one another. In such a case, the liquid metal insures the transfer of atoms from one solid substance to the other.

In the immersion of dissimilar construction materials in a liquid metal transferals of the separate substances are possible. Thus, for example, in a medium of liquid sodium the transition of the carbon in carbon steel to the surface of stainless steel is possible.

Liquid metals also appear to have a corrosive action on ceramic materials.

In the presence of a temperature gradient along a circulating system and a significant value of the solubility of some particular component of the construction material there arises a transfer of the substance in the direction of flow of the liquid metal. Such a transfer of the material of the tubes and other elements, as well as the oxides, leads to intense corrosion in individual parts and the formation of stoppages in other parts of the circulating system.

TABLE 12.1

Over-all Compilation of Data on the Resistance of Materials to Attack by Liquid Metals at 300, 600, or 800° C (neglecting mechanical hardness and heat resistance)

Liquid metal	Na-K	Li	Mg	Zn	Cd	Hg	Al	Ga	In	Tl	Sn	Pb	Bi-Pb	Bi-Pb-Sn	Sb	Bi
Melting point	-12,3 to 98,3	186	651	419,5	321	-38,8	660	29,8	156,4	303	231,9	327	125	97	630,5	271,3

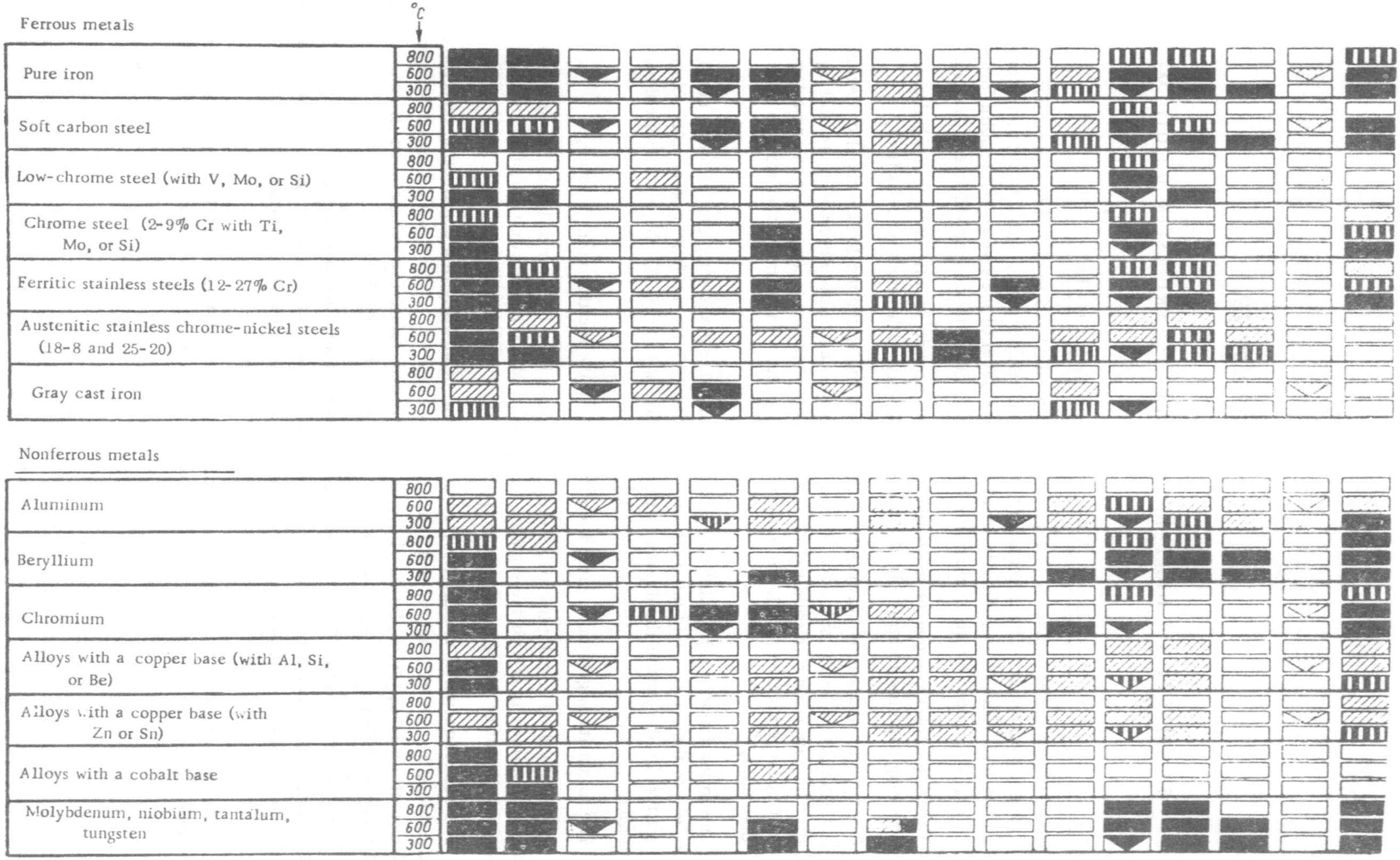

TABLE 12.1 (continued)

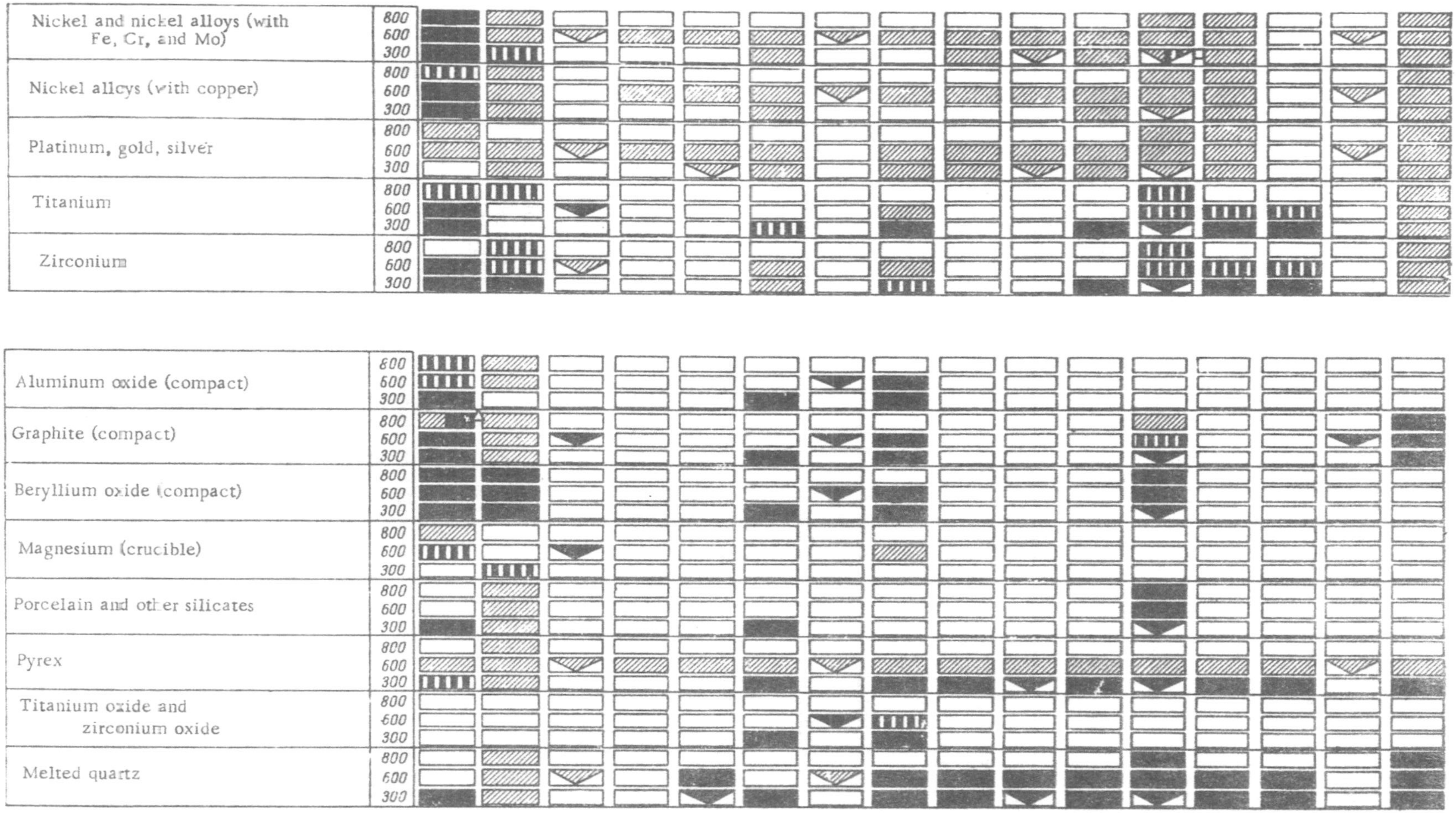

⊠ – Heat-transfer medium solidifies at the corresponding temperature above 300, 600 or 800° C

■ – Good

▥ – Limited

▨ – Poor

□ – Unknown

Δ – Graphite has poor resistance to attack by potassium and good resistance against sodium

o – **Tungsten has good resistance and molybdenum, niobium, and tantalum have poor resistance to attack by gallium**

□ – **Inconel has a limited resistance to attack by lead**

The stability of construction materials are studied under static and dynamic conditions, both with immersed and nonimmersed samples. The testing conditions, particularly the velocity of flow around the samples, their shape, and state of stress are considerably pronounced in the results obtained.

There are no sufficiently detailed data published on this problem at the present time, and only certain general information on the resistance of individual materials to attack by certain liquid-metal heat-transfer media will be presented below.

In Table 12.1 are cited data which are representative of the relative resistance of various materials to attack by liquid metals [61].

Mercury

The action of mercury on various steel under the operating conditions of power plants was studied in considerable detail by Gudtsov and Gavze [60], whose data are briefly outlined below.

TABLE 12.2
Solubility of Iron in Mercury

Solubility of iron in mercury, %	Data	Notes
< 0.01	GUI	
0.062	Nagaoki, Tamman, and Elser	
0.00134	Richards and Harrod-Thomas	
0.00007	Palmer	From 20 to 211° C almost constant
0.00001	Irwin and Russell	
$3.22 \cdot 10^{-14}$	Tamman and Coleman	
$1.5 \cdot 10^{-6}$ (at 25° C)	Marshall, Epstein, and Norton	
$1.1 \cdot 10^{-4}$ (at 700° C)	Gudtsov and Gavze	From 400 to 750° C (short-duration heating to 1000° C) a regular dependence of the solubility on the temperature and the length of testing was not observed
0.0305-0.000235 and lower *		

* The figures indicate the over-all iron content in the mercury under the conditions of the experiments. The authors do not feel that it is possible to consider them as numerical values of the solubility of iron in mercury.

The radius of the mercury atom at normal temperature is equal to 1.55 A, while the radius of the iron atom under the same conditions is equal to 1.27 A. In connection with this ratio of the atomic dimensions the formation of a solid solution of the iron with the mercury by inoculation is not possible. Nor are the conditions necessary for the formation of a solution by replacement maintained. In this connection, pure iron and mercury are almost noninteractive. Table 12.2 gives some data on the penetration of iron into mercury.

The presence of even small quantities of oxygen sharply increases the content of iron in mercury. With an increase in the flow velocity of the mercury the rate of its interaction with steel also increases considerably.

The experimental data cited in Table 12.3 makes it possible to form some idea of the interaction between mercury and various types of steel.

TABLE 12.3

Testing Results from Samples of Various Brands of Steels in a Mercury Medium*
(Testing temperature 650°, maintained for a period of 400 hours)

Brand of steel	Fraction of the sample filled with mercury	Change in wt. of the insertion piece, %	Iron content in the mercury, %	State of the surface of the sample recess and insertion piece	Remarks
20	0,08	+0,00715	0,00081	Wetted by the mercury	
20	0,2	−0,0214	0,00341	the same	
20	0,35	+0,0138	0,00359	—	
20	0,92	—	—	Wetted by the mercury	
20	0,14	—	0,00099	—	
45	0,39	+0,0492	—	Insertion piece oxidized	
U10	0,40	—	0,01475	—	
U10	0,08	0	—	—	
U10	0,9	0,0115	0,00235	Clean. Part wetted by the mercury, part covered with drops of mercury	
U10	0,22	−0,0364	—	—	The mercury was vaporized at the time of the testing
U10	0,9	—	0,00289	—	The same
15M	0,08	−0,0655	—	—	" "
15M	1,00	+0,0002	—	Clean. Wetted by the mercury	Pure mercury
15M	0,10	+0,0150	0,00161	Insertion piece oxidized	
25Kh5MA	0,08	−0,0212	—	Wetted by the mercury	
SKh8–	0,20	0	—	Insertion piece oxidized	The mercury was vaporized at the time of the testing
SKh8	0,25	+0,0103	0,00105	—	

* There was no mercury discovered in the samples, except in those cases when the surface was not coated with "skin."

TABLE 12.3 (Continued)

Brand of steel	Fraction of the sample filled with mercury	Change in wt. of the insertion piece, %	Iron content in the mercury, %	State of the surface of the sample recess and insertion piece	Remarks
SKh8	0,11	+0,0167	0,00827	–	
SKh8	0,18	—	0,00920	–	
30KhMA	0,90	+0,1590	0,00391	Insertion piece oxidized	
30KhMA	0,10	—	0,00296	Part wetted by mercury, part covered with a gray oxidation film	
30KhMA	0,89	—	0,00306	–	
30KhMA	0,19	—	0,00157	–	
EN25	0,08	—	0,00172	Wetted by mercury. The clean outer surface of the sample covered with skin	Short heating at 800° C. Diameter of the recess increased to 2.5 mm
EN25	0,22	—	—	Some spots wetted by the mercury	Short heating at 800° C. In the mercury there is a surface scaling of a magnetic nature. The diameter of sample recess increased to 3 mm.
EI181	0,08	–0,007	0,0186	Wetted by mercury	Short reheating to 900° C
EI181	0,14	+0,0210	—	Insertion piece oxidized (spots of black oxidation film)	Dark yellow film on the surface of the mercury
EI181	1,00	–0,0960	—	Coated with skin Open fracture	The mercury was vaporized. Short reheating to 1000° C.
EI454	0,08	+0,0195	0,03150	–	
EI454	0,17	–0,0456	0,00128	–	
Kh5M	1,00	0	—	–	The mercury was vaporized in the test

(Table 12.3 (Continued)

Brand of steel	Fraction of the sample filled with mercury	Change in wt. of the insertion piece, %	Iron content in the mercury, %	State of the surface of the sample recess and insertion piece	Remarks
Kh5M	0,5	−0,0072	0,00227	—	
6Kh3M	1,00	—	0,0013	—	
Kh63M	0,5	—	0,0042	Clean. Mercury droplets	On the surface of the mercury fine gray scaling, part of which is connected to the surface
Technically pure iron	0,90	−0,0036	—	Gray oxidation film. Surface of the insertion piece clean	On the surface of the mercury a light-gray film
"	0,90	−0,0175	0,0175	The same	The same
EZh2	0,90	+0,1035	—	Coated with skin	On the surface of the mercury a dark yellow film
ElaT	0,9	−0,0170	—	The same	After inspection of sample, insertion piece and sample recess were wetted by mercury
ElaT	0,9	−0,0086	—	"	The same
EZh17	0,9	−0,0167	—	"	"
EZh17 w.1% Cu	0,9	+0,0105	—	"	"
EZh17 w.2% Ni	0,9	+0,020	—	"	"
EZh17 w.2% Ni	0,9	+0,0099	—	"	"
EI257	0,9	+0,0165	—	"	"
EI268	1,0	+20,40	—	Coated with skin Open fracture	Mercury was vaporized at the time of testing
EZh268	0,9	−0,0243	—	Gray oxidation film	
EZh2	0,9	−0,033	0,0130	Clean	
EZh2	0,8	−0,4100	0,0076	"	
ElalT	1,0	−0,0039	0,0022	Clean. Partial gray oxidation film	

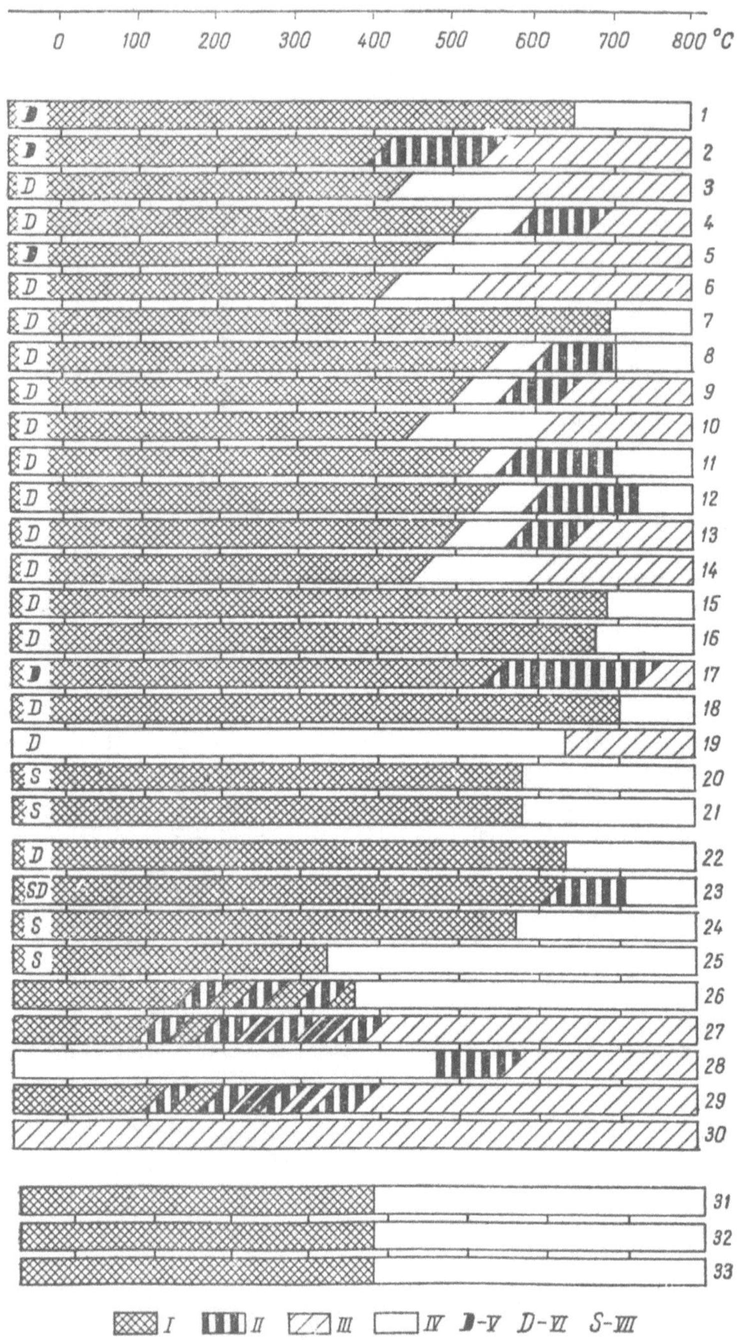

Fig. 12.1. Stability of various materials in a mercury medium at an increased temperature. I) Good; II) limited; III) poor; IV) unknown (no data available for these temperatures); V) testing in circulating mercury and in industrial melting pots; VI) testing in circulating mercury; VII) testing in mercury at rest.

Explanation of Positions

Ferrous Metals:

1) ferrous metals (Ti and Mg in Hg); 2) low-carbon steel; 3) same + 0.1 - 4% Al; 4) same + < 4% Cr; 5) steel with 5% Cr; 6) low-carbon steel + 0.5% Mo; 7) same + 20% Mo; 8) same + 1-3% Si; 9) same + 1-2% Ti; 10) same + < 2% Al + < 2% Cr; 11) Nitralloy

Explanation of Positions to Fig. 12.1. (Continued)

(1.23% Al, 1.49% Cr); 12) low-carbon steel + 5.7% Cr + 1.2% Cu; 13) same + < 4.5% Cr + < 4.5% Mo; 14) same + 5.7% Cr + 1.2% W; 15) same + 15-20% Mo + 3% Si; 16) same + 8% Cr + 5% Al + 0.3% Mo; 17) sicromo (5% Cr + 0.5% Mo + 1.5% Si); 18) low-carbon steel + 5.5% Cr + 6.4% Mo + 1.4% Si; 19) stainless steel 304 and 310 (Cr, Ni); 20) alloys with high Ni-Fe and Ni-Cr-Fe contents; 21) ferritic stainless chrome steels.

Nonferrous Metals:
22) tungsten; 23) molybdenum; 24) chromium; 25) beryllium; 26) Ta, C, Nb, Si, Ti, V; 27) Ni, Cu, and their alloys; 28) cobalt and stellites; 29) Pt, Mn, Zr; 30) Al, Bi, Cd, Ce, Au, Pb, Mg, Ag, Sn, and Zn.

Nonmetals:
31) glass; 32) ceramic; 33) graphite.

The tests do not disclose aggressive attack by the vapors of mercury on steel.

The interaction between iron and mercury at a high temperature and in the presence of oxygen proceeds according to the following scheme: the formation of an iron "skin", diffusion of the mercury through the skin layer, the formation of an intermediate layer between the surface of the iron and the skin layer.

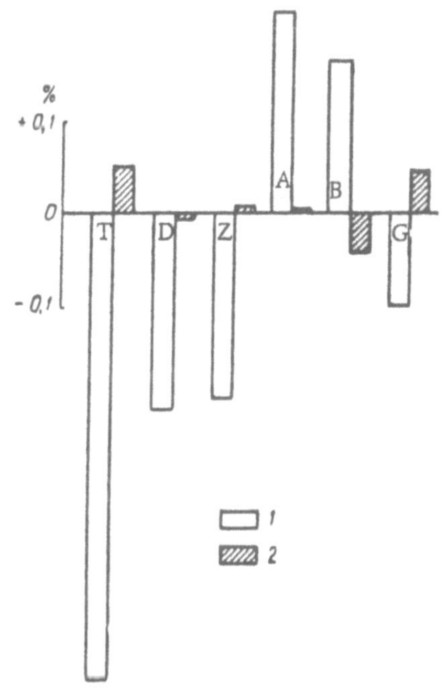

Fig. 12.2. Weight losses for steel samples tested in pure mercury (1) and mercury alloyed with magnesium and titanium (2).

The intermediate layer of mercury aids in the separation of the skin from the metal and in this way accelerates the process of formation of each successive oxidation layer. The skin, being porous, absorbs the mercury and, interspersing with the latter, forms a mass with a constant mobility.

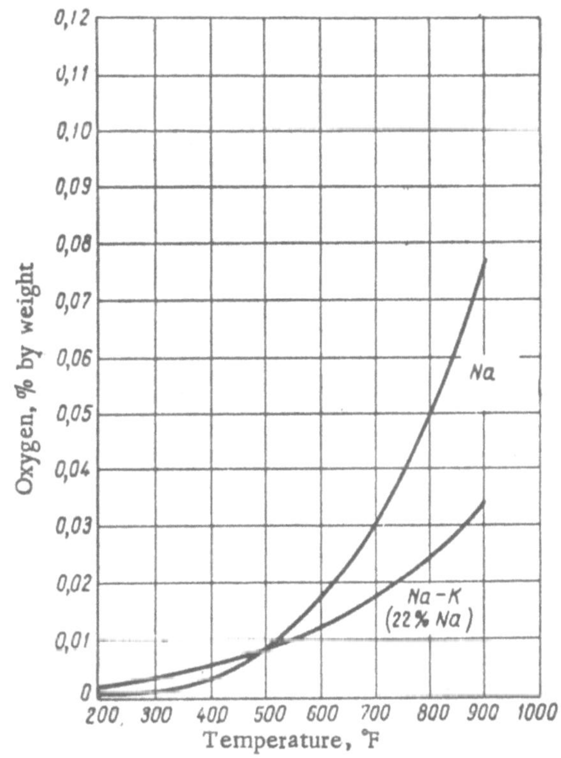

Fig. 12.3. Solubility curve of oxygen in sodium and sodium-potassium alloy as a function of temperature.

The investigations have shown that in the presence of oxygen stainless steel is more stable in mercury than noncarbon steel and medium-alloy steel.

In Fig. 12.1, are shown the results of testing with various materials in circulating mercury [61].

The aggressiveness of the attack by mercury on steel can be lowered by introducing alloying additives into the mercury. Thus, for example, according to the data of American researchers, the addition of titanium and magnesium (the latter serves to increase the wettability) to mercury makes it possible to reduce the mercury attack to a minimum with a titanium concentration of 10^{-5}. The magnesium concentration at t = 650° C should be approximately 20 times greater than the concentration of titanium.

In Fig. 12.2 are shown the results of testing with certain domestic brands of steels in pure mercury and mercury alloyed with titanium and magnesium.

The chemical composition of the tested steels is listed in Table 12.4.

Sodium

Just as in the case of mercury, sodium interacts very weakly with iron. In the temperature interval from 100 to 500° C the solubility of iron in sodium does not exceed 10^{-3}%.

The erosive action of sodium on construction materials is considerably less than for the heavy metals. Commercially pure sodium which has not been subjected to distillation contains small impurities of other elements which, under certain specific conditions, prove to have a harmful effect on steel (Table 12.5).

Of the metallic impurities the most aggressive are magnesium and calcium, which interact with nickel and certain other alloying components of heat-resistant steels.

The presence of oxides in the flow of liquid sodium considerably increases the rate of corrosion of the construction materials. In the literature a rapid development of corrosion effects is indicated when the content of oxygen in the sodium is greater than 0.02%. The oxide Na_2O dissolves poorly in liquid sodium and, settling in the elements of a circulation loop, it can give rise to pitting. The solubility curves of Na_2O in sodium and sodium-potassium alloy are shown in Fig. 12.3.

Pits are formed in the region of cold spots and contain about 20% Na_2O and 80% sodium.

The results of American investigations on the stability of various materials in liquid sodium and sodium-potassium alloy are shown in Tables 12.6-12.8. In these tables the following evaluation of the corrosion resistability is adopted:

Corrosion rate microns/year	Rating of material resistance
< 2	Good
2-25	Limited
> 25	Poor

More detailed information on the stability of certain heat-resistant materials in sodium (also under static conditions) are given in Table 12.9 [63].

Bismuth, Lead, and Lead-Bismuth Alloy

Bismuth, lead, and bismuth-lead alloy prove to exert a strongly aggressive corrosive and erosive attack on stainless steels. Aluminum and copper are considerably unstable in these media, both in the pure and alloyed forms. Nickel, which is one of the most important components of heat-resistant steels, is very unstable in lead and decomposes rapidly in bismuth.

Chrome-nickel steels are highly stable in eutectic lead-bismuth only when the nickel content is less than 15% and at temperatures less than 500-550° C.

In Table 12.10 the results of static tests with a number of materials in liquid bismuth and the alloy of lead with bismuth are presented [61].

TABLE 12.4
Chemical Composition of the Steels Treated in Fig. 12.2.

Nominal designation in Fig. 12.2	Chemical composition, %					Remnant impurities
	C	Mn	Si	Ni	Cr	
A	0,12	0,40	0,35	—	18,43	
B	0,11	0,53	0,46	—	18,57	
D	0,12	0,38	0,28	14,45	14,41	Nb=0,7 W=2,42 Mo=0,4
G	0,2	0,37	0,33	—	13,93	—
T	(commercial iron)					
Z	0,13	1,07	0,84	8	19,45	—

TABLE 12.5
Approximate Composition of the Impurities in Commercially Pure Sodium

Impurity	K	Ca	Fe	Mg	Si	O_2	H_2	C
Composition, %	0,01—0,05	0,01—0,02	0,001	0,001—0,2	0,001—0,2	>0,003	0,005	0,006

TABLE 12.6

Resistance of Certain Metals to Attack by Sodium and Sodium-Potassium Alloy During Static Corrosion Tests

With the exception of a few isolated cases, stainless steel equipment was used (According to data of the Knowles Laboratory, USA)

Metal	Temperature, °C	Resistance
Precipitated and filtered off (through a 5-micron mesh) sodium with an oxygen content of about 0.005-0.01%		
Pure aluminum, 99.99%	427	poor
Pure aluminum, 99.9%	500	the same
Aluminum alloys: 3S, 24S, 52S, 75S, 750	427	"
Poured aluminum bronze	427	good*
Armco iron	500	good
17-7 PH Armco stainless steel	427	good*
Beryllium (rod turned in the heated state)	500	good
Beryllium (chromed)	500	good*
Femico	500	good
Inconel X	500	the same
Monel	500	"
Alloy, 80% Ni, 20% Cr	500	good*
Nickel L	500	good
Nickel (1/4% Zr)	500	the same
Nickel (1/4% Zr in a nickel container)	500	"
Nitrided Nitralloy	500	good*
Nitrided stainless steel 347	500	the same
stainless steels: 303, 304ELC, 321, 347, 347 (+Ta) 347 (noncarbon), T347 (cyanided)	500	good
Low-chrome steels:		
1% Cr, 0.5% Mo	500	the same
1% Cr, 1% Mo, 0.25% V	500	"
2% Cr, 0.5% Mo	500	"
2.25% Cr, 1% Mo	500	"
3% Cr, 1.5% Si, 0.5% Mo	500	"
5% Cr, 0.5% Mo	500	"
5% Cr, 0.5% Mo, 0.5% Ti	500	"
5% Cr, 1.5% Si, 0.5% Mo	500	"
7% Cr, 0.5% Si, 0.5% Mo	500	"
Carbon steel (0.5% Mo)	500	"
Steel AISI-1019	500	good**
Zirconium	500	good
98% Zr, 2% Al	500	the same
98% Zr, 2% Fe	500	"
93% Zr, 7% Mo	500	"
92% Zr, 8% Cr	500	"
90% Zr, 10% Ti	500	"
* Testing was carried out in a separate stainless-steel vessel.		
** Decarburization was applied.		

TABLE 12.6 (Continued)

Metal	Temperature, °C	Resistance
Distilled sodium		
High-purity aluminum	450	good
High-purity aluminum	500	poor***
Aluminum alloy 2S	450	good
Aluminum alloy 2S	500	poor***
Stainless steel 347	450	good
Purified sodium-potassium with a probable oxygen content of less than 0.005%		
Pure aluminum, 99.99%	427	poor
Aluminum alloys: 3S, 24S, 52S, 75S, 750S	427	the same
Aluminum bronze T-224	427	good
Pressed beryllium	500	good*
Baked beryllium	500	the same
Oxygen-free copper (in a stainless-steel container)	427	*
Oxygen-free copper (in a copper container)	400	good
Durimet 20	427	the same
Inconel X	500	*
Mechanite	427	decarburization
Monel S	427	good*
Nickel L	500	good
Ni–Mn (40% Ni)	500	good
Ni–Mn (80% Ni)	500	the same
Ni-resist	427	decarburization
Nitrided stainless steel 347	500	good
Tantung-Ni (70% Ni)	500	good*
Tantung: A 166, A 171, A 192, A 166 spring	500	the same
Titanium	500	*
Vanadium	500	good****
Zirconium	427	good*
Zirconium	500	good
Stainless steel: 302, 304, 310, (25% Cr 20% Ni), 316, 321, 347, 35-15-Ni-Cr	500	the same
Stainless steel 416	427	good*
Stainless steel 440 C	427	the same
Steel 52100	427	*
High-speed tool steel 6-6-2	427	*
High-speed tool steel 18-4-1	427	*

* Testing was carried out in a separate stainless-steel vessel.
*** The plane sample was placed at a distance of $^1/_{16}$ inch from a stainless-steel plate.
**** Testing was carried out in a separate nickel vessel.

TABLE 12.6 (Continued)

Metal	Temperature, °C	Resistance
Sodium, purified by the addition of potassium in the container		
Pressed beryllium	500	good
Baked beryllium	500	the same
Nickel A (with a 2% addition of Ca)	500	limited
Nickel A (with a 10% addition of Ca)	500	poor
Nickel L (with a 0.1% addition of Ca)	500	good
Stainless steel 347	500	the same
Na–K (44% Na), filtered through a 5-micron mesh		
High-purity aluminum	450	good
Aluminum alloy 2S	450	the same
Stainless steel 347	450	good

TABLE 12.7
Resistance of Cemented Carbides to the Action of Precipitated and Filtered Sodium*

Nominal composition	Resistance	
	500°C (932° F)	750°C (1382° F)
30% Ni+8% $NbTiTaC_3$+62% TiC	Good	Limited
20% Ni+15% $NbTiTaC_3$+55% TiC	Poor	—
20% Ni+8% $NbTiTaC_3$+72% TiC	Good	Good
20% Co+80% TiC	The same	—
20% Co+15% $NbTiTaC_3$+65% TiC	"	—
5% Co+15% $NbTiTaC_3$+80% TiC	Limited	—
100% TiC	Good	Good
Cemented on cobalt:		The same
20% Fe+80% TiC	The same	"
10% Co+8% $NbTiTaC_3$+82% TiC	"	Limited
30% Ni+70% TiC	"	The same
20% Ni+80% TiC	"	—
20% Ni+15% $NbTiTaC_3$+55% TiC	"	—

* The initial weight of the samples repeatedly tested in sodium at 750° C is assumed equal to the final weight of the samples after testing at a temperature of 500° C. This provides a source of some insignificant error.

TABLE 12.8

Resistance of Certain Ceramic Materials to Attack by Precipitated and Filtered Sodium

Ceramic material (composition in moles, unless otherwise stated)	Resistance	
	500° C (932° F)	750° C (1382° F)
Synthetic magnesium oxide crystals	Good	Good
Beryllium oxide (BeO)	–	–
Stabilized	Limited	–
$1BeO + 1\ Al_2O_3$	Good	–
$1MgO + 4ZrO_2$	–	Poor
$1Mg + 8BeO + 1Al_2O_3$	Good	Limited
$160BeO + 2Al_2O_3 + 1ThO_2 + 2\%\ CaO$	Limited	–
$76BeO + 4Al_2O_3 + 20ZrO_2 + 2\%\ CaO$	Good	Good
$3MgO + 90BeO + 1ZrO_2$	–	Limited
Hot-pressed synthetic mica	Decomposition	–
TiN	Cracking	–
TiC_2	Decomposition	–
$CaZrO_3$	Poor	–
TiO	Decomposition	–
ZnO	–	–
TiC	Cracking	–
Be_2C (after testing for swelling due to hydration)	Limited	–
$MoSi_2$	Decomposition	–
B_4C, TaC, TiC, VC, ZrC	Good	–
Silicon carbide (SiC) crystals	The same	Limited
Mixture of the carbides B_4C and SiC	Limited	–
Mixture of carbon and titanium borides	Good	–
Zirconium boride ZrB_2	Limited	–
Calcium boride CaB_6	Poor	–
Zirconium nitride ZrN	Good	–
Beryllium nitride Be_3N_2	Limited	–
Titanium nitride TiN	Poor	–
Beryllium oxide BeO	Good	–
Chrysoberyl $BeO{-}Al_2O_3$	The same	–
Spinel (hot-pressed) $MgO{-}Al_2O_3$	Limited	–
Zirconium with CaO	The same	–

TABLE 12.9
Corrosion of Heat-Resistant Materials in Molten Sodium Under Static Conditions

Material	Results	Temperature of the sodium, °C	Test duration, hours	Change in wt of the sample, %	External appearance
Molybdenum (99.9%)	does not corrode	940	168	0,0001	Lustrous
	pitting to a depth of 0.025 mm	1500	100	—	"
Tungsten (99.9%)	almost no corrosion	900	168	from −0,07 to +0,01	Almost no sign of corrosion
Tantalum (99.9%)	The same	900	168	from −0,09 to +0,01	In some cases becomes dull
Cannametal (KI38A) TiC —80%, Co—15%, WC, TaC—5%	Light corrosion	900	168	from −0,07 to +0,20	No change
AWG-brand graphite	Corrodes	900	168	−6,0	Surface is corroded
AUF-brand graphite	Heavy corrosion	900	841	−60,0	Clear signs of corrosion
Silicon carbide single crystal	Some indications of corrosion	900	168	+1,0	No change
Synthetic spinel ($MgO \cdot 3{,}5Al_2O_3$)	Completely dissolves	925	168	−100,0	Disappears

TABLE 12.9 (Continued)

Material	Results	Temperature of the sodium, °C	Test duration, hours	Change in wt. of the sample, %	External appearance
Morganite RR aluminum oxide, porosity 6%	Heavy corrosion	940	168	—	Crumbles into a number of soft black grains
Aluminum oxide (synthetic sapphire)	Very slight in comparison with morganite	900	168	−1,0	Remains transparent
Magnesium oxide, porosity 12% (LM—160—35)	Some indications of corrosion	940	168	+1,2	Blackens and swells
Magnesium oxide (single crystal)	Almost no corrosion	925	168	−0,02	Remains transparent
Zirconium oxide, porosity 23% (Norton LZO—169B—35; CaO stable)	Considerable corrosion; samples disintegrated	890	168	+3,7	Blackens and swells
Molybdenum disilicide (poured)	Evidently slight corrosion occurs	900	168	—	Lustrous
Thorium oxide (Norton LT—10A—05)	Almost no corrosion	925	168	−0,06	No change

TABLE 12.10

Corrosion of Heat-Resistant Materials in Liquid Bismuth Under Static Conditions

Material	Result of tests	Temperature of the bismuth, °C	Duration, hours	Content in the bismuth, %	Metallographic or visual observation
Molybdenum (99.9 %)	Evidently no corrosion	1000	167	0,06 Mo	Wetted by the bismuth
Tungsten (99.9 %)	No clear signs of corrosion	1000	167	0,001 W	—
Tantalum (99.9%)	Highly pronounced intercrystalline breakdown	1000	227	0,11 Ta	Wetted by the bismuth
Rhenium, baked from powder in a graphite crucible	Essentially no corrosion	1000	250	0,006 Re	No wetting
Cannametal (K138A TiC—80%, CO—15% WC, TaC—5%)	No clear indications of corrosion	1000	188	0,01 Ti, 0,02 Co	The same
AWG-brand graphite	The same	1400	168	—	"
AUF-brand graphite	"	1400	168	—	"
Silicon carbide (Norton bar)	Indications of corrosion	1000	227	0,17 SiC	Intergranular penetration

TABLE 12.10 (Continued)

Material	Result of tests	Temperature of the bismuth, °C	Duration, hours	Content in the bismuth, %	Metallographic or visual observation
Aluminum oxide	No clear indications of corrosion	1000	150	0,0002 Al	No marked changes
Morganite RA-1232	Light corrosion	1345	2,5	0,08 Al	The same
Magnesium oxide (Norton bar)	Very slight corrosion	1000	227	0,01 Mg	Changed from light yellow to clear-white
Zirconium oxide	Essentially no corrosion	1000	167	0,003	Turned dark-gray
Molybdenum disilicide (bar, obtained by hot-pressing)	No evident signs of corrosion	1400	168	—	According to the data of metallographic investigations, intercrystalline corrosion is not observed
Thorium oxide ThO_2 (Norton bar)	Light corrosion	1000	167	0,06 Th	Changed from light-gray to clear-white
Tantalum carbide (from Brewer, University of California)	No indications of corrosion whatever	1000	167	0,001 TaC	
Wickor (sample tubes at a temperature of 1000° C)	No clear indications of corrosion	1000	167	0,002 Si	No marked changes. Remains transparent

Chapter 13

INSTRUMENTATION

Temperature Measurements

The fundamental temperature-measuring devices for liquid-metal heat-transfer media are resistance thermometers and thermocouples. The first are suitable for measuring the mean flow temperature. Ordinary measurement principles are used.

The choice of material for thermocouples and resistance thermometers is determined by the level of the temperature measurements. These same factors determine the type of insulation as well. The instrumentation should be suitably protected against direct contact with the heat-transfer medium, which gives rise to corrosion and electrical short-circuiting. In this connection, soldering the thermocouple junction to a protective casing is not recommended. The protective casing should be prepared from a corrosion-free material and, insofar as possible, should have a low inertia. For temperature measurements up to 500-600° C the protective casing can be made from thin chrome and nickel-chrome tubing.

For the researcher's needs thermocouples made from wires having diameters of 0.1-0.3 mm and inserted in steel tubes having outside diameters of 1.5-2 mm are used. For purposes of insulation, hollandizing — a coating of manganate and other materials which are resistant to attack in the high-temperature region — is applied.

In the assembly of such thermocouples impairment in the insulation is possible, so that it is necessary to have a control, which, for example, can be effected by measuring the electrical resistance of the thermocouple.

For measurements of very high temperatures a tungsten wire resistance thermometer can be used.

The response of a thermocouple and protective casing having a diameter of about 1.5 mm amounts to approximately 1 sec in time lag.

The materials of the thermocouples and resistance thermometers should be resistant to the action of neutron radiation.

The welding of thermocouples (which should be very thin) to the tubes and other construction elements is usually accomplished by fusion-welding equipment. In experimental investigations the proper placement of thermocouples on a heat-exchange surface is a decisive factor in the accuracy of the measurements.

According to one method the thermocouple can be laid into the wall of the tube through a face drilling. The thermocouple can also be laid in a groove and covered over with the metal of the tube. This seal can be accomplished with the help of a metal spray gun, galvanizer, or by other methods applicable to the particular metal of the heating surface.

In measuring the mean flow temperature a resistance thermometer or thermocouple can be introduced into the flow by means of a sleeve pipe.

In measuring small differences in temperature, an arrangement of differential thermocouples is desirable. A differential thermocouple measures the temperature deviation at a given point from the temperature (measured by an ordinary thermocouple) at some initial reference point.

For such a method of measurement the possible error of 1-2° C in the determination of the absolute temperature does not have too great an influence on the difference measurement.

Pressure Measurements

At room temperature mercury and sodium-potassium alloys are in the liquid state. Therefore, for measuring the pressure in systems which contain these heat-transfer media ordinary manometers with Bourdon tubes are applied. For mercury the Bourdon tube should be made of steel. Here, both the sensitive probe (pick-up) and the pulse line should be filled, as usual, with the cold heat-transfer medium.

Insertion of the Bourdon gage by the usual scheme for a system with high-melting metals ($t_m > 100°$ C) does not guarantee reliability in the measurements, since the heating necessary in this case, up to temperatures exceeding the melting point of the heat-transfer medium, gives rise to considerable error in the measurements.

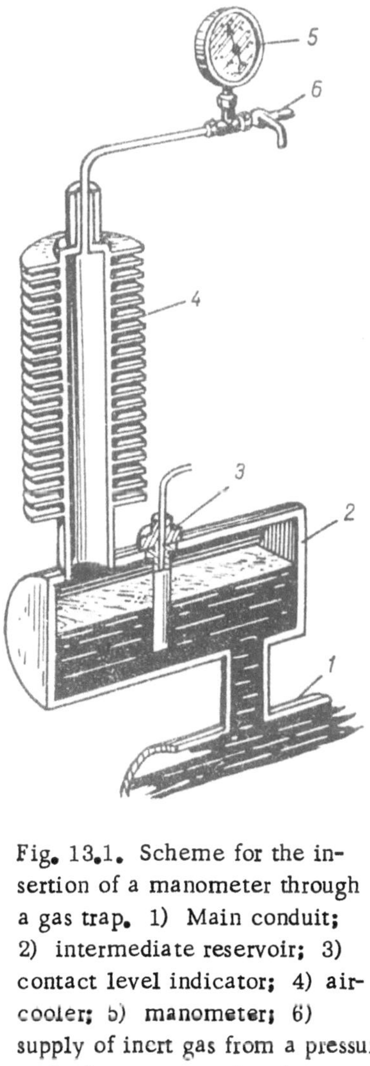

Fig. 13.1. Scheme for the insertion of a manometer through a gas trap. 1) Main conduit; 2) intermediate reservoir; 3) contact level indicator; 4) air-cooler; 5) manometer; 6) supply of inert gas from a pressure tank, for compensating the pressure in the main line.

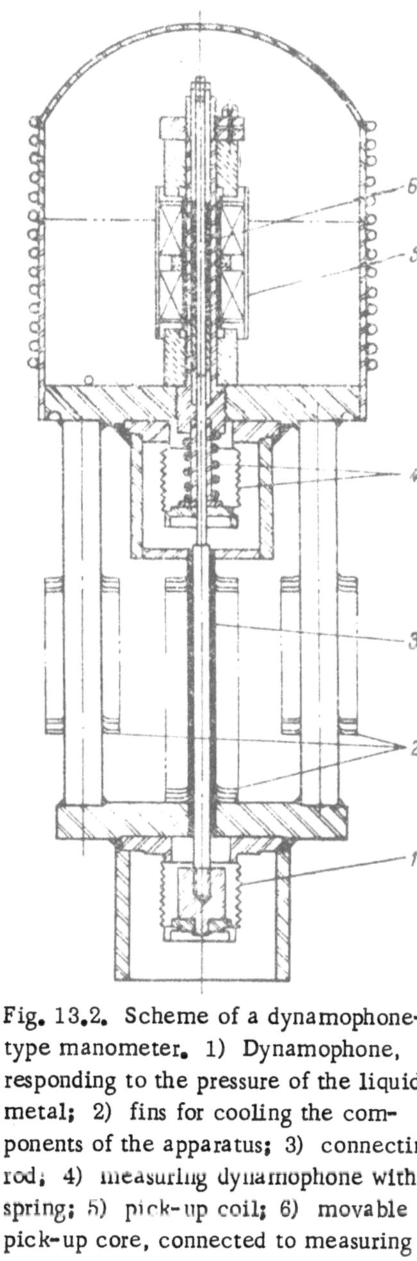

Fig. 13.2. Scheme of a dynamophone-type manometer. 1) Dynamophone, responding to the pressure of the liquid metal; 2) fins for cooling the components of the apparatus; 3) connecting rod; 4) measuring dynamophone with spring; 5) pick-up coil; 6) movable pick-up core, connected to measuring dynamophone.

In laboratory practice it is possible to apply ordinary manometers, which are inserted through a gas trap, in systems which are under careful surveillance. One such arrangement is shown schematically in Fig. 13.1.

Special types of manometers, which can be applied in direct contact with the liquid metal, have membranes or dynamophones with pneumatic or electrical pick-ups for their instrumentation. The latter is preferable in view of its higher response and the fact that it has no need for an air column.

In Fig. 13.2, the scheme for a manometer of the dynamophone type with an inductive pick-up is shown. A hermetically sealed dynamophone is immersed in the liquid metal and by means of a rod is joined with another dynamophone, located in the normal temperature zone. The second (measuring) dynamophone is provided with a spring. The soft-iron core of a differential transformer is attached to the measuring dynamophone. The measuring part of the apparatus is protected against high temperature by means of cooling fins. The characteristics of the dynamophones and springs are selected so as to reduce the action of thermal expansion and other factors on the calibration of the apparatus to a minimum.

The application of high-temperature deformation meters as pick-ups gives good results.

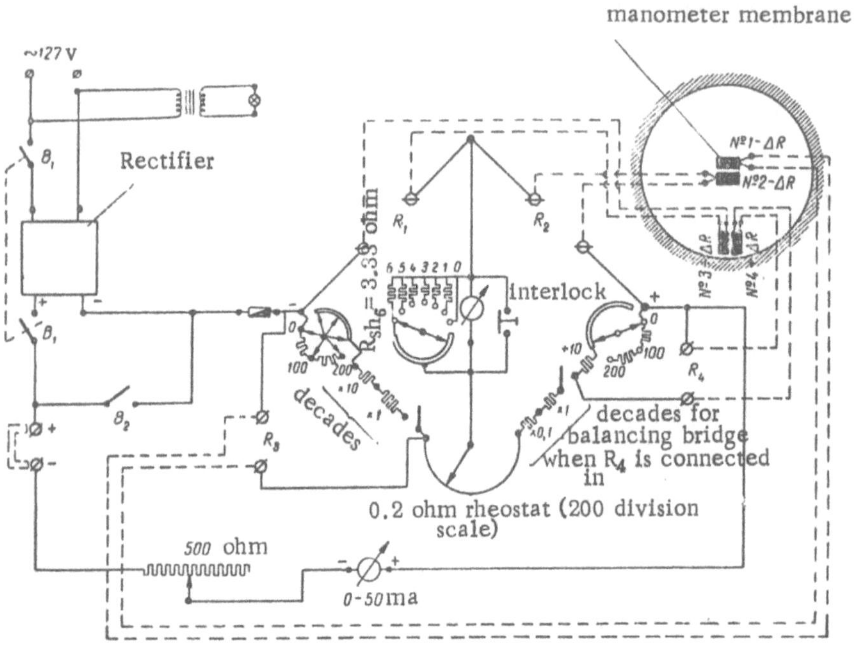

Fig. 13.3. Principal electrical circuit of a deformation-measuring manometer with four pick-ups.

In Fig. 13.3, a circuit for the inclusion of four deformation-meter pick-ups into a measuring bridge is shown. The deformation meters are placed on the membrane of the manometer in such a way that two of them operate during stretching, two during constriction. In doing this, temperature compensation takes place and with this paired connection of the deformation meters their calibration remains almost invariant over a wide temperature range. With the inclusion of four (two pairs) of deformation meters the sensitivity of the apparatus is augmented. The membrane or dynamophone can be directly immersed in the flow or connected by a short sensing line to the main line. In the latter case the apparatus is provided with an electric heater with automatic temperature regulation. The remote apparatus makes it possible for the membrane to operate at temperatures lower than those in the main line, and in this way a smaller variation in the elastic properties of the instrumentation is insured.

In Fig. 13.4 is shown a photograph of one of the experimental deformation-measuring manometers heated and an ordinary-type manometer. The calibration curves of such a manometer are straight lines, while the sensitivity of the apparatus varies with the current variation (Fig. 13.5).

The reliability of deformation-meter pick-ups to a large extent depends on the material of the wire and the heat-resistant element used to attach the deformation meter to the membrane. The characteristics of these materials should remain almost constant for a period of several thousand hours.

If the apparatus is subjected to the action of a neutron flux, the materials from which it is made should be resistant under these conditions. In addition to the deformation-meter pick-ups described, the application of pick-ups of the capacitative and other types is possible.

Fig. 13.4. Deformation-measuring manometer with a large membrane, shown with an ordinary manometer.

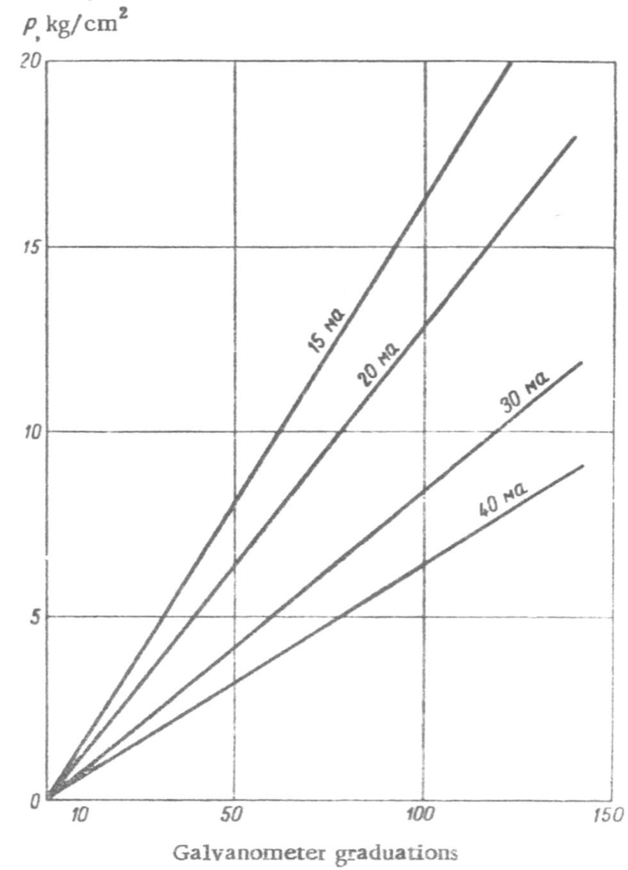

Fig. 13.5. Calibration curves of the deformation-measuring manometer.

The most precise measurement of the mass flow rate of a liquid-metal heat-transfer medium is a standard metering valve with a differential manometer, in which the working medium is the liquid metal itself. The buffer medium in such a manometer is fulfilled by a neutral gas, supplied to it under a pressure corresponding to the pressure in the main line. In measuring the mass flow rates of mercury and sodium-potassium alloy the manometer is not heated. In the case of mercury an incompressible fluid (specifically, water) can serve as the buffer medium, which simplifies the utilization of the apparatus considerably. In using this procedure for the flow rates of other liquid-metal heat-transfer media, the differential manometer should be heated.

In Fig. 13.6, a scheme for measuring the mass flow rate of a liquid metal by means of a metering valve and differential manometer (piezometer) is shown. In Fig. 13.7, the construction of one type of heated differential manometer is shown.

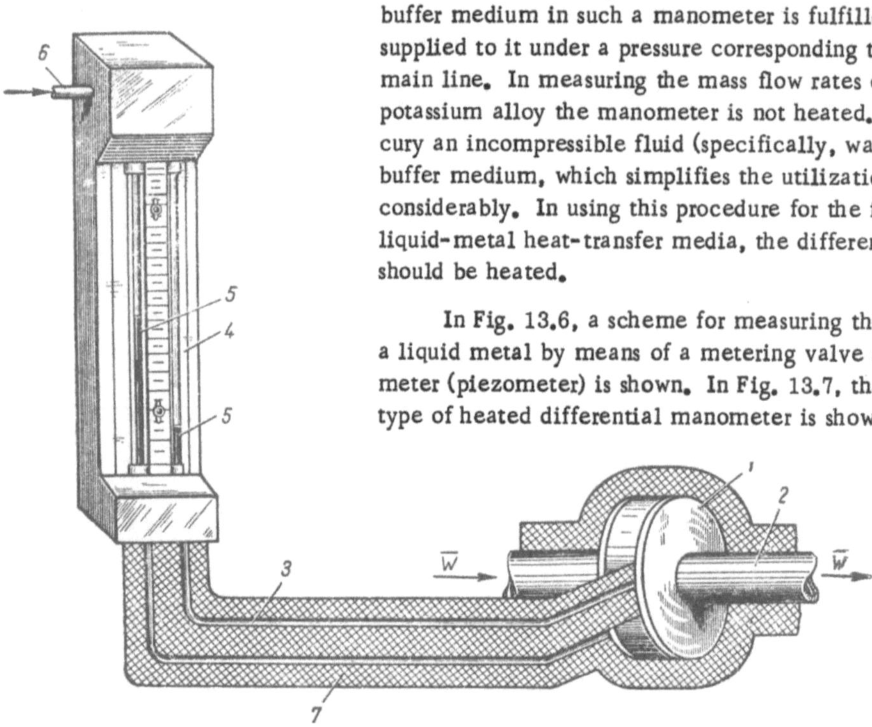

Fig. 13.6. Scheme for the laboratory measurement of the mass flow rate of a liquid metal by means of a metering valve and differential manometer. 1) Valve; 2) main line of flow; 3) connecting tube; 4) manometer housing, heated internally; 5) piezometric tube; 6) connecting branch for conducting the inert gas to the upper manometer block; 7) insulation.

In addition to the standard metering valve, a Venturi tube can be applied. The use of fine plates or washers is unsuitable because of the possibility of erosion wear of their edges and a resultant alteration in the calibration.

The application of rotameters is possible with the use of induction- or capacitance-type pick-ups. Here, the shape of the operating part of the rotameter should be sufficiently smooth, in order to avoid erosive attack by the liquid metal.

Under performance conditions the application of a metering valve with a differential manometer is unsatisfactory for all liquid-metal media, with the exception of mercury. In this case a closed differential manometer, in which there are no glass tubes, should be applied. A reliable type of apparatus along this line is an induction differential manometer.

For measuring the mass flow rates of other liquid-metal heat-transfer media under performance conditions and in a number of experimental conditions the magnetic flowmeter is used almost exclusively. This apparatus insures the complete hermeticity of the flow contour, it has no moving or sealed parts, its dimensions are relatively small, and it has almost linear characteristics.* The principle of its construction is shown in Fig. 13.8. The conduit is flanged (which, of course, is not mandatory) and is inserted between the poles of a permanent magnet. On opposite sides of the conduit contacts leading to a millivoltmeter are soldered or inserted directly into the flow. The liquid metal, being a conductor, distorts the magnetic lines of force, so that an electric current is developed in the circuit. The voltage of the latter is directly proportional to the flow velocity of the metal, i.e., to its mass flow rate.

* In the Soviet Union the first practical development of electromagnetic flowmeters was made by Vel'tishchev.

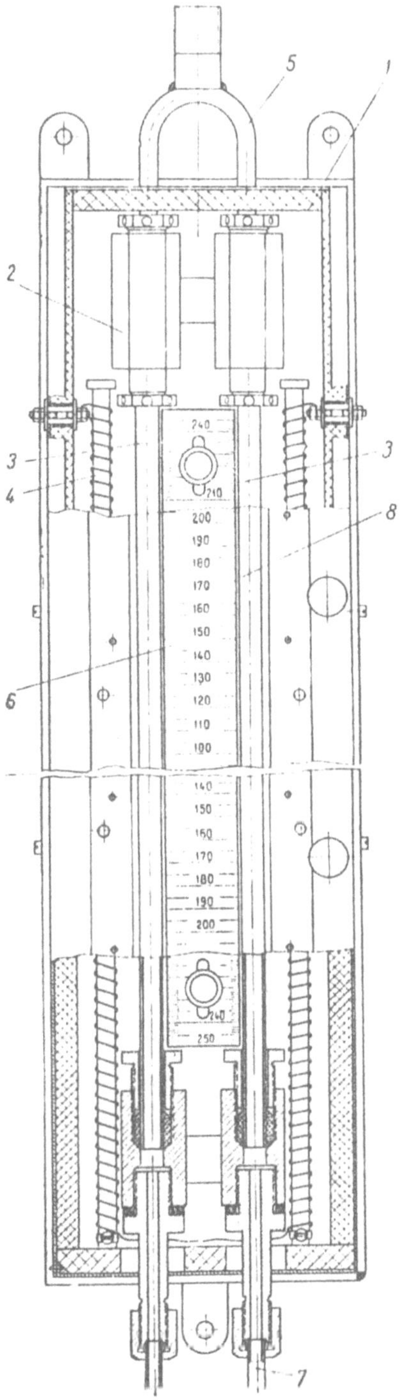

Fig. 13.7. Construction of a heated differential manometer.

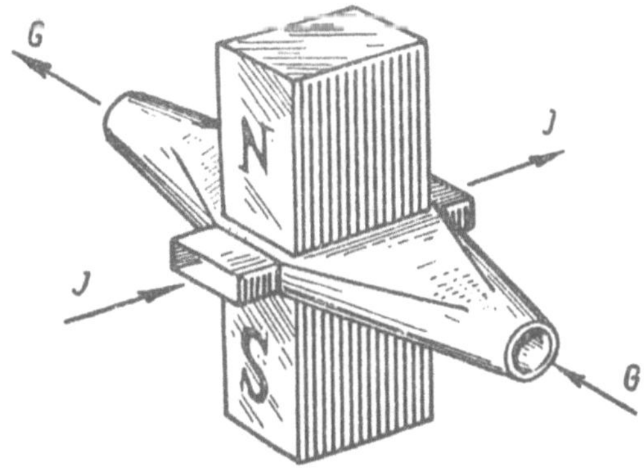

Fig. 13.8. Schematic drawing of a magnetic flowmeter. N, S) the poles of the magnet; G is the conduit; J is the electrical current.

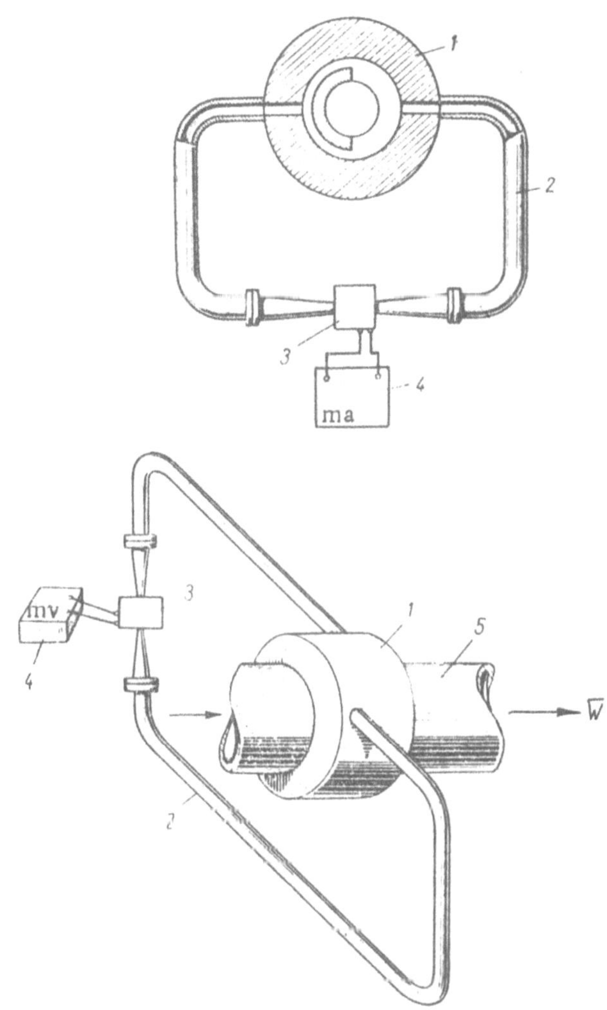

Fig. 13.9. Scheme for the by-pass connection of a magnetic flowmeter. 1) Valve; 2) tubing; 3) magnetic flowmeter; 4) potentiometer; 5) main line of flow.

The electrical conductivity of the walls of the tube of the magnetic flowmeter should be almost the same as or less than the conductivity of the liquid metal.

The theory of this apparatus is a special case of the theory of electromagnetic pumps [79].

In the measurement of mass flow rates through small-diameter tubes these need not be flanged. In the measurement of mass flow rates through tubes of larger dimensions the application of the apparatus shown schematically in Fig. 13.8 requires magnets of large dimensions. Moreover, the characteristic of the apparatus deviates from a linear one, which in all probability is connected with the attenuation of the magnetic field by the circling current. In this case it is expedient to apply a standard magnetic flowmeter with a pipe of small diameter, which, in a manner similar to the differential manometer, is connected to the sensing lines of a standard metering valve. The scheme for such an apparatus is shown in Fig. 13.9. Under the influence of the pressure drop elicited by the metering valve, a small part of the liquid metal flows through the tube of the magnetic flowmeter. The velocity of the liquid flowing through the magnetic flowmeter is determined by the pressure head at the metering valve and by the hydraulic resistance of the sensing line according to the equation

$$w_f = \sqrt{\frac{2g\Delta p_V}{\gamma\zeta_f}}, \tag{13.1}$$

where Δp_V is the pressure drop between the reference momenta in the valve, γ is the specific gravity of the metal flowing through the tube of the apparatus, ζ_f is the coefficient of hydraulic resistance of the sensing tubes and the tube of the apparatus (sensing line) and g is the gravitational acceleration.

The head at the valve in turn is equal to

$$\Delta p_V = \zeta_V \frac{\gamma w_c^2}{2g} \tag{13.2}$$

where ζ_V is the nominal coefficient of hydraulic resistance of the valve, referred to the pressures at the reference points for the momenta in the valve, and w_c is the flow rate of the metal in the conduit.

Consequently, the mass flow rate of the liquid-metal heat-transfer medium through the conduit is determined from the equation

$$G = \gamma w_c = \gamma w_f \sqrt{\frac{\zeta_f}{\zeta_V}}. \tag{13.3}$$

Since the quantity $\sqrt{\frac{\zeta_f}{\zeta_V}}$ is usually almost constant, the characteristic of such an insertion of the magnetic flowmeter (i.e., the dependence of the emf of the apparatus on the flow rate of the metal through the valve) is linear.

As a measuring standard it is convenient to use a magnetic flowmeter with a small-diameter tube. The dimensions of the valves should depend on the diameter of the conduit.

Experiment shows that the most suitable method is the insertion of the contacts of the magnetic flowmeter directly into the liquid-metal flow. In this event the contacts are electrically insulated from the tube by a corrosion-resistant material. The contacts themselves should be made from a material which is stable in the given medium, and they should not move out far from the wall of the tube, in which case they would fall into the region of high flow velocities.

The presence of oxides and other contaminating agents in the heat-transfer medium influences the calibration of the apparatus and can cause it to vary with time.

Measurement of the emf of the magnetic flowmeter is usually performed by a potentiometer method, which insures the utmost accuracy and compensation for the effect of the electrical resistance of the circuit. Under performance conditions it is expedient to employ recording potentiometers.

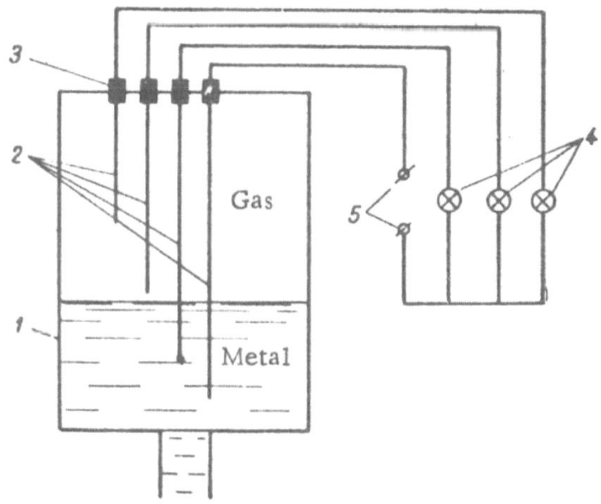

Fig. 13.10. Schematic drawing of a contact level indicator. 1) Reservoir; 2) contacts; 3) insulators; 4) signal lamps; 5) current sources.

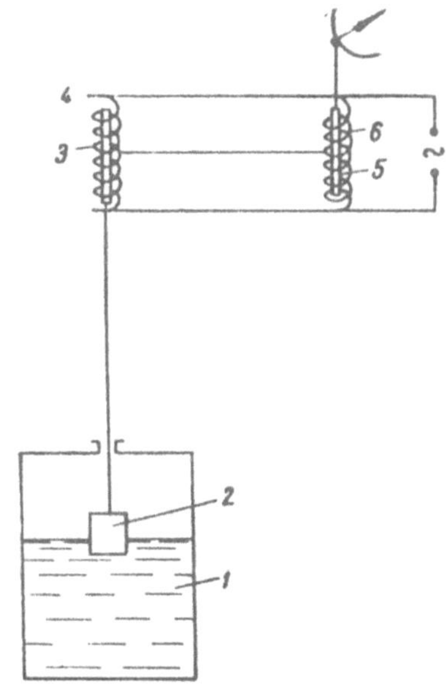

Fig. 13.11. Schematic drawing of a float level indicator with induction pick-up. 1) Reservoir; 2) float; 3) pick-up core; 4) pick-up coil; 5) instrument core; 6) instrument coil.

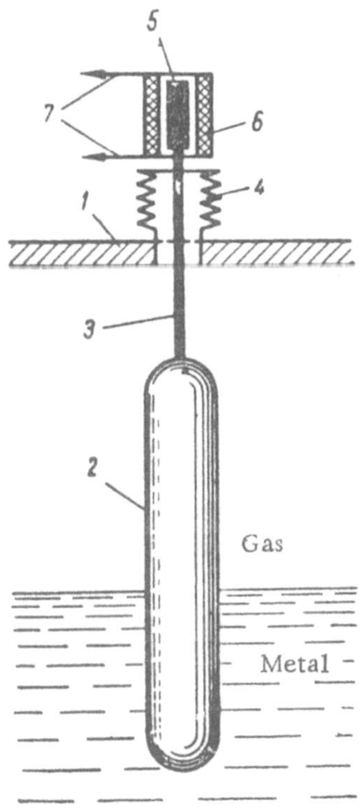

Fig. 13.12. Schematic drawing of a float instrument with a dynamophone transmitter. 1) Top of the reservoir; 2) inertial float; 3) float rod, attached to the dynamophone; 4) dynamophone, which balances the force of the hydraulic pressure on the float; 5) transmitter core; 6) transmitter coil; 7) wire leading to instrumentation circuit.

Level Measurements

A fully reliable method for measuring the level of a liquid metal is that of a system of contacts placed at various heights (Fig. 13.10). A signal appears upon shorting of the circuit at the instant of contact of the level of the liquid metal with one of the contacts. The shortcoming of this method is the absence of a continuous reading.

Continuous level measurements can be conducted by measuring the variation in the resistance of a rod submerged into a reservoir with the liquid metal in it. As the level of the liquid metal increases, part of the rod is inundated and its resistance is lowered. However, the calibration of such an instrument can change from sticking of the liquid metal and oxides on the exposed part of the resistance.

A schematic drawing of a level indicator with a fundamental working part in the form of a float and with an inductive pick-up is shown in Fig. 13.11. The response is transmitted to standard instrumentation. In this case, the movement of the float coincides with the magnitude of the variation in level, which necessitates large dimensions for the pick-up coil.

In Fig. 13.12, a schematic drawing of a similar apparatus with a long float, which acts on a dynamophone, is shown. The dynamophone shifts the transmitter coil. In such a scheme the displacement of the level of the liquid is associated with a smaller displacement of the dynamophone and a correspondingly smaller movement in the transmitter coil. The last factor results in normal dimensions and a good characteristic for the pick-up.

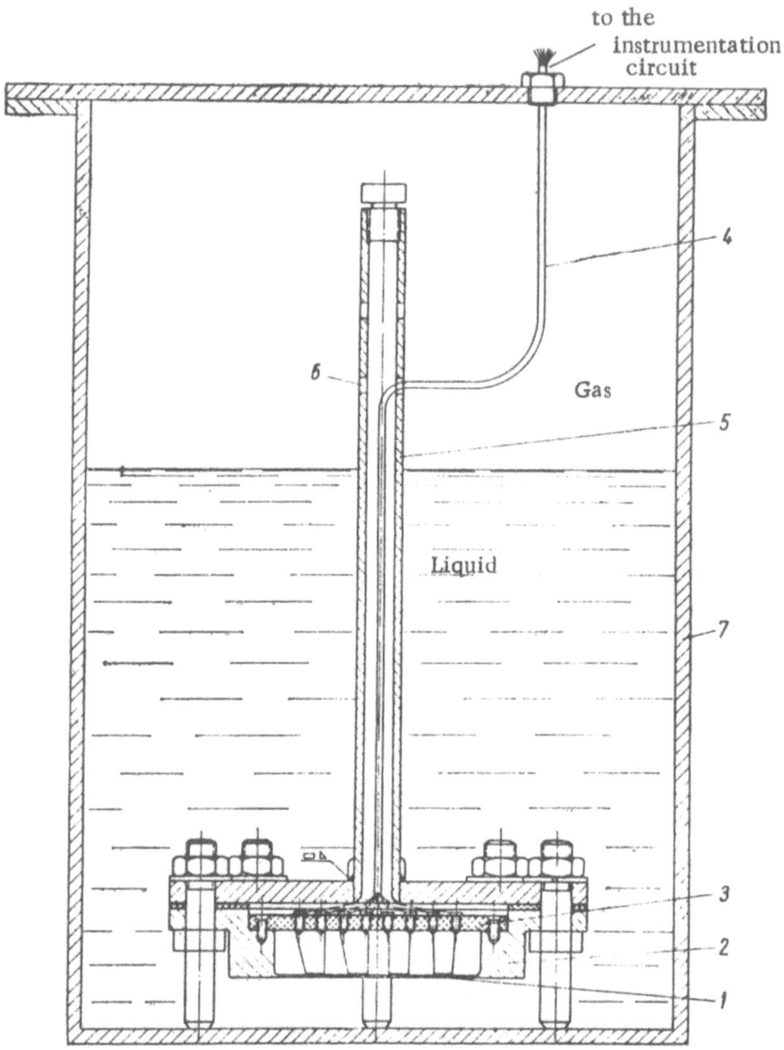

Fig. 13.13. Schematic drawing of a deformation-measuring level indicator. 1) Membrane with deformation meters; 2) housing; 3) contact block; 4) wire leading to instrumentation; 5) tube; 6) opening; 7) reservoir.

Fig. 13.13. schematically illustrates a deformation-measuring level indicator with the same instrumentation as in the case of the deformation-measuring manometer described earlier.

LITERATURE CITED

[1] M. A. Styrikovich, I. E. Semenovker, A. R. Sorin, Sovetskoe Kotloturbostroenie 9, 316 (1940).

[2] M. A. Stryrikovich and I. E. Semenovker, J. Tech. Phys. (USSR) 10, 16, 1324 (1940).

[3] A. N. Lozhkin and A. A. Kanaev, Binary Heat Engines [In Russian] (Mashgiz, Leningrad, 1946).

[4] A. A. Kanaev, Kotloturbostroenie 2, 18 (1953).

[5] V. M. Borishanskii and S. S. Kutateladze, Energomashinostroenie 6 (1957).

[6] S. S. Kutateladze, Fundamentals of Heat-Exchange Theory [In Russian] (Mashgiz, Leningrad, 1957).

[7] M. A. Mikheev, V. A. Baum, K. D. Voskresenskii, and O. S. Fedynskii, Coll.: The Design and Theory of Reactors [In Russian] (Acad. Sci. USSR Press, 1955), p. 139.

[8] K. D. Voskresenskii, Proceedings of the International Conference on the Peaceful Uses of Atomic Energy 9 (UN, New York, 1956).

[9] M. A. Mikheev, Heat Transfer Principles [In Russian] (Gosenergoizdat, Moscow, 1956).

[10] R. C. Martinelli, Trans. Am. Soc. Mech. Engrs. 69, 8, 947 (1947).

[11] R. N. Lyon, Chem. Eng. Prog. 47, 2, 75 (1951).

[12] H. E. Brown, B. H. Amsted and B. E. Short, Trans. Am. Soc. Mech. Engrs. 79, 2, 279 (1957).

[13] B. Rightmire and H. Taylor, J. Iron and Steel Inst. 175, 167 (1953).

[14] M. I. Korneev, Teploenergetika 7, 30 (1955).

[15] D. English and T. Barrett, General Discussion of Heat Transfer (Inst. Mech. Eng., 1951), p. 458.

[16] T. C. Doody and A. H. Younger, Chem. Eng. Prog. Symposium Series 5, 33 (1953).

[17] S. E. Isakoff and T. B. Drew, General Discussion of Heat Transfer (1951), p. 405.

[18] H. A. Johnson, W. J. Clabaugh, and J. P. Hartnett, Trans. Am. Soc. Mech. Engrs. 75, 6, 1191 (1953); 76, 4, 505 (1954).

[19] E. R. Gilliland, R. J. Musser and W. R. Page, General Discussion of Heat Transfer (1951), p. 405.

[20] L. M. Trefethen, General Discussion of Heat Transfer (1951), p. 436.

[21] B. Lubarsky and S. J. Kaufman, "Review of experimental investigation of liquid metals," Heat Transfer Report of the National Advisory Committee on Aeronautics 1270 (1956).

[22] L. I. Novikov, A. N. Solov'ev, E. M. Khabakhpasheva, V. A. Gruzdev, A. I. Pridantsev, and M. Ia. Vasenina, Atomnaia Energiia I, 4, 92 (1956).

[23] W. B. Hall and A. E. Jenkins, J. Nuclear Energy 1, 244 (1955).

[24] R.A.Tidball, Chem. Eng. Prog. Symposium Series 5, 233 (1953).

[25] R. D. Brooks and A. L. Rosenblatt, Mech. Eng. 75, 5, 363 (1953).

[26] T. Trocki and D. E. Nelson, Mech. Eng. 75, 6, 472 (1952).

[27] Stuart McLain, Reactor Engineering Lectures (Chicago, 1955).

[28] H. B. Squire, in the book: Problems of Modem Aerodynamics, edited by L. Howarth (Russian translation) (IL, 1955), Vol. 1.

[29] E. R. Eckert, Introduction to the Transfer of Heat and Mass (McGraw-Hill, New York, 1950).

[30] M. V. Kirichev, M. A. Mikheev and L. S. Eigenson, Heat Transfer [In Russian] (Gostekhizdat, 1940).

[31] S. C. Hyman, C.F. Bonilla, and S. W. Ehrlich, Chem. Eng. Prog. Symposium Series 49, 5, 21 (1953).

[32] W. H. McAdams, Heat Transmission, 3rd Edition (New York, 1954).

[33] V. M. Borishanskii and S. S. Kutateladze, J. Tech. Phys. (USSR) 28, 4 (1958).

[34] V. K. Semenchenko, Surface Effects in Metals and Alloys [In Russian] (Gostekhizdat, Moscow, 1957).

[35] A. I. Lozhkin and V. I. Krol', J. Tech. Phys. (USSR) 8, 21, 1872 (1938).

[36] D. I. Gremilov, Coll.: Combination Power Engines and Cycles [In Russian], Trudy TsKTI, Book 23 (Mashgiz, Leningrad, 1952), p. 44.

[37] M. I. Korneev, Teploenergetika 4, 44 (1955).

[38] P. O. Siryi, A. A. Kanaev and A. A. Lozhkin, Sovetskoe Kotloturbostroenie 8-9, 375 (1938).

[39] E. K. Averin and G. N. Kruzhilin, Izv. AN SSSR Otdel Tekh. Nauk 10, 131 (1955).

[40] S. S. Kutateladze, Heat Transfer During Boiling and Condensation [In Russian] (Mashgiz, Leningrad, 1949, 1952).

[41] V. M. Borishanskii and V. K. Migai, Energomashinostroenie 6 (1958).

[42] V. M. Deriugin, Energomashinostroenie 4, 30 (1958).

[43] V. M. Borishanskii, S. S. Kutateladze, E. A. Kozadova, et al., Coll.: Heat-Exchange Problems Arising Upon Changing of the Aggregate State of Matter [In Russian] (Gosenergoizdat, Leningrad, 1953).

[44] H. C. Schwenk and R. H. Shannon, Power 99, 11, 92 (1955).

[45] Handbook on Liquid Metals, Supplement (Washington, 1955).

[46] R. N. Lyon, A. S. Faust and A. L. Katz, Chem. Eng. Prog. Symposium Series 51, 17, 41 (1955).

[47] W. B. Hall and T. I. M. Crofts,Proc. Inst. Mech. Engrs. (London) 170, 10 (1956).

[48] G. Gröber and S. Erk, Die Grundgesetze der Wärmeübertragung (Russian translation) (ONTI, Moscow, 1936).

[49] A. V. Lykov, Theory of Heat Conduction [In Russian] (Gostekhizdat, Moscow, 1952).

[50] Coll.: Problems of Turbulence [In Russian] (ONTI, Moscow, 1936).

[51] N. E. Zhukovskii, Papers, Vol. 1 [In Russian] (OGIZ, 1949).

[52] O. J. Kleppa, J. Chem. Phys. 18 (1950).

[53] V. Jacobson, J. Chem. Phys. 20 (1952).

[54] S. S. Kutateladze, J. Tech. Phys. (USSR) 28, 4 (1958).

[55] R. A. Seban and T. T. Shimazaki, Trans. Am. Soc. Mech. Engrs. 72, 6, 803 (1950).

[56] G. N. Kruzhilin, J. Tech. Phys. (USSR) 6, 9, 1578 (1935).

[57] A. N. Solov'ev, Atomnaia Energiia III, 12, 550 (1957).

[58] A. M. Sperrow and J. L. Gregg, Trans. Am. Soc. Mech. Engrs. 2 (1956); 78, 435 (1956).

[59] Thermal Calculation of the Boiler Aggregate (Normative Method),[In Russian] TsKTI, VTI (Gosenergoizdat, Moscow, 1957).

[60] N. T. Gudtsov and N. N. Gavze, The Action of Mercury as a Heat-Exchanging Medium on Steel in Power Plants [In Russian] (Acad. Sci. USSR Press, Moscow, 1956).

[61] Materials of the Atomic Energy Commission (USA), Nuclear Reactors, Vol. II: Technology of Nuclear Reactors (Russian translation) (IL, Moscow, 1957).

[62] M. Sittig, Sodium, Its Manufacture, Properties and Uses (New York, 1956).

[63] E. Reed, J. Am. Ceram. Soc. 37, 3, 146 (1954).

[64] Handbook on Liquid Metals, Supplement (Washington, 1950).

[65] Landolt, Physikalische Chemische Tabellen (Berlin, 1927-1937).

[66] Tiessen, Scheel, Seel, Wiss. Abhandl. Phys. Techn. Reichs 2, 73 (1895); 3, 1 (1900).

[67] Cailletet, Colardean, Riviere, Compt. rend. 130 (1900).

[68] P. D. Borisov, Sovetskoe Kotloturbostroenie 11 (1938).

[69] G. Sauerwald and K. Tepler, Z. anorg. u. allgem. Chem. 157, 117 (1926).

[70] L. F. Golubev and V. A. Petrov, Zavodskaia Laboratoriia 7 (1938).

[71] S. V. Sergeev and E. V. Poliak, Materials of the Council on the Viscosity of Liquids and Colloidal Solutions [In Russian] (Acad. Sci. USSR Press, 1945).

[72] Kh. Kalilov, J. Tech. Phys. (USSR) 8, 13-14, 1249 (1938).

[73] O. A. Saunders, Proc. Roy. Soc. (London) 72, 948 (1939).

[74] O. S. Fedynskii, Coll.: Heat Transfer and Thermal Simulation [In Russian] (Acad. Sci. USSR Press, Moscow, 1958).

[75] N. S. Kondrat'ev, Coll.: Heat Transfer and Thermal Simulation [In Russian] (Acad. Sci. USSR Press, Moscow, 1958).

[76] M. A. Mikheev, O. S. Fedynskii, V. M. Deriugin and V. I. Petrov, Coll.: Heat Transfer and Thermal Simulation [In Russian] (Acad. Sci. USSR Press, Moscow, 1958).

[77] N. A. Nikol'skii, Coll.: Heat Transfer and Thermal Simulation [In Russian] (Acad. Sci. USSR Press, Moscow, 1958).

[78] Piercy and Wihni, Phil. Mag. 16 (1933).

[79] Applied Magneto-Hydrodynamics [In Russian] Report of the Physics Institute of the Academy of Sciences of the Latvian SSR VIII (1956).

[80] K. D. Voskresenskii and E. S. Turilin, Coll.: Heat Transfer and Thermal Simulation [In Russian] (Acad. Sci. USSR Press, 1958).

[81] L. I. Gel'man, Teploenergetika 3 (1958).

MIX
Papier aus verantwortungsvollen Quellen
Paper from responsible sources
FSC® C105338

FSC
www.fsc.org

If you have any concerns about our products,
you can contact us on
ProductSafety@springernature.com

In case Publisher is established outside the EU,
the EU authorized representative is:
Springer Nature Customer Service Center GmbH
Europaplatz 3, 69115 Heidelberg, Germany

Printed by Libri Plureos GmbH
in Hamburg, Germany